The Rajah of Darjeeling Organic Tea

Makaibari

Rajah Banerjee

Delhi • Bangalore • Mumbai • Kolkata • Chennai • Hyderabad

Published by

Cambridge University Press India Pvt. Ltd.

under the Foundation Books imprint

Cambridge House, 4381/4 Ansari Road, Daryaganj, New Delhi 110 002

Cambridge University Press India Pvt. Ltd.

C-22, C-Block, Brigade M.M., K.R. Road, Jayanagar, **Bangalore** 560 070

Plot No. 80, Service Industries, Shirvane, Sector-1, Nerul, **Navi Mumbai** 400 706

10 Raja Subodh Mullick Square, 2nd Floor, **Kolkata** 700 013

21/1 (New No. 49), 1st Floor, Model School Road, Thousand Lights, **Chennai** 600 006

House No. 3-5-874/6/4, (Near Apollo Hospital), Hyderguda, **Hyderabad** 500 029

© Cambridge University Press India Pvt. Ltd.

Photos and Sketches © Author

First Published 2008

ISBN: 978-81-7596-605-5

All rights reserved. No reproduction of any part may take place without the written permission of Cambridge University Press India Pvt. Ltd., subject to statutory exception and to the provision of relevant collective licensing agreements.

Published by Manas Saikia for Cambridge University Press India Pvt. Ltd. and printed at Sanat Printers, Kundli.

Contents

Foreword .... *v*

Acknowledgements .... *vii*

1. The Beginning 1
2. Creating an Organic Makaibari 12
3. Setting an Example 34
4. An Eventful Year 44
5. The Community 60
6. Makaibari Tea 70
7. The Tea Deva 80
8. Lore of the Logo 86
9. Spreading the Spirit of Makaibari 94
10. Makaibari's Wildlife 110
11. Makaibari Fables 118
12. Through the Visitors' Eyes 130

Epilogue *164*

Index *167*

Foreword

As far as first impressions go, your first meeting with Rajah (Swaraj Kumar) Banerjee is unequivocally unforgettable. Within moments of entering his office-cum-study, with its trove of relics from years of visitors, Rajah will offer two things – some of the world's finest tea and the most intense conversation. Not long after you settle into your chair, he might quiz you on the four uses of mulch, enlighten you on the magic of biodynamic agriculture and ascertain the true purpose of your visit to his Makaibari Tea Estate. He will challenge your perceptions and quite probably change your mind on at least one thing. You will rise from your chair refreshed by tea yet dazed by brilliance. You will replay the meeting in your mind for days to come.

This is part of the magic and mystique of Rajah Banerjee.

Darjeeling Black Orthodox Tea is renowned worldwide and Rajah Banerjee is a living legend in Darjeeling. He is a champion of the organic tea movement, a social activist for tea labourers and small organic farmers, a tireless anthropologist preserving the cultural heritage of the Himalayan region and an environmentalist fighting to conserve its rich biodiversity.

Rajah is one of the few owners living in his tea estate, an estate that has been with his family for more than 150 years. Living in the Makaibari Tea Estate enables Rajah to manage its day-to-day operations that result in some of the world's finest teas. His hands-on approach also lends to improving the quality of life of the tea estate residents, as he is unique in understanding the relationship between the need for vibrant communities in which the tea labourers live and the quality of tea that is produced. He is constantly challenging himself to create better living conditions for Makaibari residents through alternative livelihood programmes and improvements in the social infrastructure.

I became acquainted with Rajah through the Community Health Advancement Initiative (CHAI), a corporate social responsibility project funded by the Tazo Tea Company and Mercy Corps. Through CHAI, he administratively and financially supports alternative livelihoods for the unemployed on Makaibari Tea Estate – constructing tourist lodges owned and managed by residents, paying market prices to women-headed self-help groups who manufacture handmade-paper tea boxes, and promoting small private farmers who produce organic tea as well as medicinal and aromatic plants. He works tirelessly to improve the social fabric of the villages in the estate through educational, cultural and community health programmes; he has built schools, potable water projects, community centres, and a library.

Rajah Banerjee, the idealist, is a prime mover in the social transformation of Darjeeling's poorest and most underserved families. However, it is Rajah Banerjee, the visionary, that will stand the test of time. Rajah is a follower of Rudolf Steiner, the Austrian philosopher who founded the biodynamic agriculture movement, the predecessor to organic farming. Based on Steiner's biodynamic principles, Rajah converted his tea estate to an organic tea estate in 1988, abjuring the accepted practice of maximising yields through artificial fertilisers and pesticides. It has taken almost two decades for the majority of owners to realise the long-term benefits of sustainable organic agriculture.

Now Rajah is promoting the spread of organic farming throughout Darjeeling. He argues that small private tea growers are the future of Darjeeling Tea. He believes that, for the health of the industry, the large tea estates should eventually break up – including his – to make way for the small organic farmer. This radical thinking is a threat to the barons of the tea industry. Still, Rajah instigates with a wry smile and eyes focussed on the future. The social activist in Steiner would be proud of his student.

This book is important not only because it documents the remarkable achievements of one man, but also because it is about the future of Darjeeling's environment and tea industry.

Rajah did not become the proprietor of one of the most famous tea estates in the world by working from a posh air-conditioned office in Kolkata. He is breathing the Himalayan air with his employees to nurture his tea bushes. Rajah moves on foot and horseback, interacting and learning from everyone who crosses his path, from the lone traveller, tea estate residents, social activists, visiting scholars, and agronomists to name a few.

This book gives a rare glimpse of one of Darjeeling's greatest characters, the Thunderbolt Rajah. Imagine yourself sitting across the desk from him. Take time to savour the tea and thoughts.

John Strickland
Regional Director
Mercy Corps.
Darjeeling, India
15 October, 2007

Acknowledgements

Joe Smillie – co-author of *Soul of Soil* and Makaibari's first inspector. A titan, not only physically but organically.

Walter Rudert – a treasure house of knowledge from whom I have learnt humility; one of the finest beings in the biodynamic movement.

Kiran Tawadey – owner of Hampstead Tea and Coffee Company, UK – who was the catalyst in making Makaibari organic.

Horst Michaelson and Helmut Kaufmann of OTG (Ostfriesische Tee Gesellschaft), Germany – they were my inspiration to march alone boldly with organic farming in 1988, as avid Makaibari tea buyers.

The Ishii family from Tokyo – who love all the sustainable impulses at Makaibari.

Anupa Mueller – owner of Eco-Prima Inc., USA – who gave up her highly paid corporate job to throw in her lot with Makaibari.

Sworup Dutt – my friend, philosopher and guide, who has been by my side at my darkest hour of need.

Srirupa – for spearheading the Fairtrade philosophy at Makaibari and abroad.

Sujit and Archita Chakravarty – for sharing the healing powers of Makaibari holistically with every being.

Professor Faltin and Thomas Rauchle of PWST (Projektwerkstatt Teekampagne), Germany – who support the Makaibari way unconditionally.

Michel Finkoff, Paris, France – who has steadfastly voyaged the gamut of Makaibari's sustainable journey.

Abhro Aich – my dear friend, for being ever obliging.

Debra Kellner – who flew out of the cosmos to share her radiance beyond the Silver Tips Imperial.

John, Sanjay, Sunil, Ruebens and all at CHAI, Darjeeling – for their unmitigated cooperation in making the mantra of 'partnership not ownership' a reality.

To the greatest modern day philosopher, Rudolf Steiner and all denizens of the Makaibari community – my heartfelt gratitude for providing me with the answers to the three questions that assail all of us: Where do we come from? What are we doing here? Where do we go?

Rajah Banerjee
Owner, Makaibari Tea Estate

THE BEGINNING

Captain Samler looked around him. The military outpost at Tetuliya, though outwardly spick and span, was two-tiered. The British officers enjoyed all the trappings of conquering rulers, while the natives, impeccable in khaki and Gurkha hats, were herded in barrack dormitories, adequately nourished with bland rice and lentils, and were bullied to execute all the dirty work – from menial drudgery to ferocious fighting in the battlefield. He bore this stoically for five years and then something snapped. One moonlit evening, he broke free. Together with ten Gurkha sepoys, he raided the armoury and decamped.

The group headed for the hills, 241 kilometres north. It took them a week to arrive at an unknown spot – densely wooded and teeming with wild animals, birds and insects. This was at the start of the monsoon, so they cleared a patch and planted maize. It was to be a future food source, and would also provide another amazing use, which they were to discover in the near future.

Makaibari – the land of the dancing mist

The first batch of military police wended their way up to the maize patch when the corn was ripening. The Samler community, forewarned by their bustle, lay in wait inside the maize field. The unsuspecting troops puffed their way up to the clearing, and the very next moment, they were in full flight, as a volley of musket fire sailed over their heads. Two or three more unsuccessful sorties resulted in lowering troop morale to such an extent that Samler was left alone to pursue his own destiny. The road that he had hacked with his band was later to be named Pankhabari, and of course, the outpost where he built his homestead bore the legend 'Makaibari' (cornfield). This was in 1835.

Meanwhile, Dr Campbell, with the blessings of the British Raj, opened tea nurseries at Kurseong and Darjeeling in the 1840s. Samler, who believed in the spirit rather than the letter of law, quietly pinched a few tea saplings and brought the first commercial tea plot in the district, into being. Nobody troubled him, but no official recognition was forthcoming either – not that it bothered him in the least. To date, this first tea plot is called 'Paila Khety' (the first plantation).

A hundred kilometres from Kolkata, a young lad living on an affluent zamindari landholding had a dream. By the age of fourteen, he had mastered both English and French. He wanted to sail to London to become a barrister. Though born and raised in the lap of luxury, he was sensitive to the poverty and injustice around him. He wanted to redress it in his own way. Stating his case lucidly and confidently to his older brother, the patriarch of the principality their family controlled, only resulted in a virulent scolding from his guardian. Early the next day, he jumped on his favourite horse from the stables and fled with a few possessions. A fortnight later, he arrived at Tetuliya – weak, hungry and dishevelled.

The British Commandant noticed the unusual fire burning in the stripling's eyes and offered him refuge. A week later, Girish Chandra Banerjee (G.C.) was the unofficial correspondent for the entire

garrison, on account of his command on English and his wonderful writing. Two years later, at the ripe age of sixteen, he had cornered the entire pony express service, and effectively, the communication for the region. The area boomed, and G.C. was canny to seize the prime locations at Kurseong (land of the white orchid) and Darjeeling (land of the thunderbolt). At twenty, he was the wealthiest man in the region. However, he was shrewd enough to underplay this wealth by living simply and assisting the needy. Samler and G.C. inevitably became good friends. In the late 1850s, as Samler lay ailing, he summoned the young G.C. and bequeathed him Makaibari's official registration in 1859. This was the beginning of the miracle.

The young G.C. expanded dramatically, servicing the needs of a rapidly growing town and tea community. He had neither the time nor the inclination to nurture the demands of a nascent tea plantation. Instead, he turned the plantation over to a British managing agency, while he consolidated his holdings. When he breathed his last in 1898, the Banerjees and their wealth were indeed an enviable sore point with the colonial masters.

Tara Pada (T.P.), his eighteen-year-old son, an undergraduate at the prestigious Presidency College, was thrust at the helm of the empire. Both prospered. Meanwhile, his father's dream to be a barrister was realised by his younger brother, when he qualified as a brilliant barrister. The sibling's oratorical skills propelled him to the upper echelons of Kolkata society. He had the city under his feet, as he was feted and sought by kings and commoners alike. The unprecedented attention went straight to his head. He ventured on a lifestyle, unimaginable even to potentates. Maintaining his harem (which had the most attractive and expensive courtesans from various continents) and hosting non-stop lavish parties, finally took their toll on both his purse and his health. He was compelled to seek his share from T.P. Wisely, T.P. agreed to a split heavily in his brother's favour, knowing full well that anything short of it would have ended in a full-scale legal battle with no winners. The younger brother happily accepted and promptly launched himself on a course of self-destruction.

Darjeeling – land of the thunderbolt

Shortly thereafter, T.P. annulled the managing agency and placed his younger son, Pasupati Nath (P.N.), at the helm of the business in 1939. P.N. was a man with remarkable reflexes, eyesight and courage. He was not interested in tea, his big passion being *shikar* (hunting). Bagging the big cats or the Bengal tiger was his sole aim. Honing his talents with a rifle and shotgun, he became the scourge of all creatures of the forest, and the envy of all the British planters. His extraordinary marksmanship demanded his presence at all hunts – whether at princely states or planters' shoots. These were the most prestigious events in the social calendar.

Once while hunting at Makaibari, P.N. was caught in a dramatic thunderstorm. He observed that the run-off from the woods was crystal clear, whereas the gushing water from the plantation was brackish. It did not take him long to realise that the wealth of the tea plantation – the topsoil – was being washed away. It was apparent to him that the fallen leaves from the trees provided a barrier, preventing soil erosion in heavy downpours. The challenge for him was to develop a similar blanket for the tea plantation. Armed with his realisation, P.N. approached his father. The patriarch was brutally abrupt in his response, 'I'm not here to make matters easier. It is your discovery, you find the solution'. P.N. came away from Happy Valley, where his father resided, very unhappy indeed.

Mr Crees was a legendary planter who managed his own garden, Lopchu. On learning of P.N.'s dilemma, Mr Crees asked P.N. to meet him at seven o'clock the following morning. P.N. arrived sharply on time, riding on his horse through torrential monsoon rain. Together, they set off to trek the entirety of Lopchu. They did not even take a break for lunch. Mr Crees nibbled on his sandwiches as they marched, while P.N. could only eye them hungrily. He forced himself to turn his thoughts away from his rumbling belly, realising it was a test. They plodded on. Finally, at five o'clock in the evening, after a ten-hour walk in pouring rain, Mr Crees stopped abruptly and asked, 'How do you feel, young Banerjee?' P.N.'s mind raced. His answer had to be correct, or the mantra would not be forthcoming. He replied, 'I have been up and about since three this morning. I have not eaten all day, and we have been tramping around in the rain for hours. If I had been at Makaibari, I would have been exhausted. Here, I am exhilarated. The ground underfoot is as soft as a cushion.'

As he spoke, the last vestige of his weariness vanished, as the answer revealed itself. P.N. gazed in wonder at the acres of various shrubs and grasses that Mr Crees had so lovingly planted at regular intervals between tea. The workers had to simply take the loppings, and spread them as a blanket between the rows of tea. He hugged Mr Crees joyously. Guru and disciple set off arm-in-arm for a celebratory drink. What a miracle it was that turned P.N. from a hunter, to a protector of the soil. In 1945, with mulching, Makaibari embarked on her miraculous organic voyage.

rows of tea

An Indian undergraduate's existence at the University of London could only teeter between two poles. Traditionally, he would be content to meekly potter around in his dingy bed-sit, rarely venturing out to savour the wickedness of the Western world. But life for him upon returning home after the mandatory stint was initially euphoric. In the rounds of endless parties

(where he would be the centre of attraction) this timid soul would transform to a veritable oratorical tiger. He preened about, bloated with his own self-importance. To prolong this blissful state, he would invent stories. He would claim that he had prowled through the campus in London, conquering legions of emancipated beauties. Alas, before the deluded dreamer could descend from his fantasy world, a sharp mother-in-law (the real barracuda) would snare the victim for her nubile daughter. The high flier was then quietly assimilated into the tedium of work, mandatory golf, procreation and the stereotyped social milieu.

At the other extreme were individuals who valued their self-respect. An Indian in Britain was a second-class citizen. The reason was two-fold – first, the colonial legacy of the vanquished and second, the glorious distinction of the built-in tan. An Indian student was totally traumatised on first setting foot on English soil. Discrimination was rampant at all levels, be it housing, being served at a store, travelling or vacation job hunting. However, the moment one began to take stock, solutions to overcoming these barriers appeared. One learned to be a reflector. Then the forces at play became easy waves to ride on and he began to have fun. The vistas were innumerable for one's positive growth. The range of options that then presented themselves would have been inconceivable in India. As an undergraduate in London, I was enjoying life to the lees. Under these circumstances, there was no incentive for me to return to the family plantation at Makaibari.

In 1970, after a four-year sojourn, I returned home for a holiday. P.N., my father, astutely realised that any pressure would only hasten my flight

young Swaraj Kumar Banerjee

from the coop. Instead, he cleverly lulled me into a false sense of security. 'Son, you are living on the edge. Relax, ride, hunt, have a good time. Return refreshed and recharged.'

I was delighted to try my hands on the pheasants with a pair of matched Purdy's shotguns, and to careen around the steep, narrow and precipitous plantation roads on the champion thoroughbred, both artfully gifted to me by my father. However, I had no intentions of remaining on the plantations – or so I thought. A month, at most, of this pleasant interlude and back I would return to the bustle and excitement of London's concrete jungle, where I had successfully charted my course for the present, and which promised endless material goals for the future. There was no way I would remain at Makaibari!

The twenty-first day of August in 1970 proved to be momentous. Riding home in the afternoon, Invitation, my prized horse, shied when a wild boar ran across his path. I was thrown off the horse. In the split second that I fell, I perceived a brilliant band of white light, connecting me to the trees in the forests around me. The woods sang out melancholically in an incredible concerto, 'Save us! Save us!'. The moment was eternal, and would alter the course

magic of Makaibari

of my life forever. It was pure magic – a miraculous experience beyond time and space. I hit the ground with a thud, and soon a cluster of nearby pluckers was on me, enquiring about my well-being, dusting and picking me up.

I thought to myself, 'Gosh, I've been trying to buy love and recognition thousands of miles away, and all of it was under my very nose, for free.' There and then, the die was cast. I would put my heart and soul into saving the vanishing woodlands and liberating the people of Makaibari. I had no clue how to go about it – only the inspiration of the recent vision, strengthened further by the love and concern of the workers, who had not existed in my world, till that very instant. That evening at supper, my parents' joy was unbounded when they heard my decision.

chemicals – the chain destroyers

CREATING AN ORGANIC MAKAIBARI

Makaibari spans 670 hectares over six separate ridges. The tea covers 270 hectares, while the woodlands cover twice that area. The tea is ensconced between the forests. This makes supervision difficult. In other plantations, the tea is a contiguous carpet. This means that one has to work thrice as hard to cultivate tea at Makaibari.

I devoured all the available literature as well as scientific publications on tea. My readings revealed that the concept of mulching was only touched upon and there was no reference to mulching in any published journal. Nature did not require any external help to sustain and evolve the myriad life forms comprising the ecosystem at Makaibari. How otherwise, could so many varieties of trees exist, cheek in jowl in a sub-tropical rain forest and harbour the wide diversity of organisms that they did? Human beings only imbalanced nature for their own use, and needed science to bolster the equation of their monstrosity. Science, it seemed, desired victory over nature, while conveniently forgetting that the creature that wins against the environment ultimately destroys itself.

the organic tea garden

One day, as I walked in a distant part of the plantation, I found a pheasant, which had succumbed to insecticide poisoning. Close observation for over a month witnessed the demise of a wild cat that fed on it. Finally, the carcass of a panther completed the chain of destruction. This was shocking. Everyone is aware of the trauma of Nagasaki and Hiroshima, yet farmers all around the world continue to bomb their lands. I pleaded my case for banning all spraying and chemical applications at Makaibari to P.N. passionately. He listened, but never acted. It was depressing. I perceived that although the mulch created wonderful topsoil, all the millions of organisms were annihilated by a single dose of fertiliser. Why create ideal conditions for organisms to flourish only to destroy them? Discussing this issue with P.N. only resulted in alleviating my misery. P.N. would nod his head in agreement, but not move a finger to address the issue.

After a month, I had an idea. There was a patch of tea on a steep slope in the deep woods. Nobody visited this spot, not even P.N. I took two of the oldest workers on the estate into my confidence, and we manured the area secretly at night. It took the three of us, almost a month to carry the compost from a nearby village and spread it on our site. The tea leaves harvested from this site were processed separately. During the daily tasting sessions, P.N. commented on the exceptional quality of this tea. However, each time he probed about the source area of the tea, I was able to cleverly deflect his curiosity. I secretly carried a Dictaphone during these sessions for an entire year. Finally, when I revealed the experiment to him, P.N. supported me wholeheartedly; even providing cows to the workers and teaching them composting techniques. Urea to cow dung manure was a definitive conversion.

Shortly thereafter, a high-powered delegation of tea experts visited Makaibari. I tagged along with this group. Arriving at a precipitous spot where only tall thatch grew, one of the luminaries commented that no tea could ever grow there. I promptly countered that nothing was impossible. All eyes were on me in the stunned silence that followed. I boldly wagered that tea could be successfully cultivated, even under seemingly impossible conditions. The bet was

laid, and three years later, the group assembled to witness the wonder of one of the finest patches of tea. In good humour, the group christened it as 'S.K.'s folly'. Love and care could transform anything.

The plantation ladies were the most hardworking workers on the estate. They were up at four o'clock in the morning every day to gather firewood and prepare meals for the family. They cleaned and scrubbed, and they sent the children off to school. All before seven o'clock sharp, when they joined their friends to pluck tea. An hour's break at midday was followed by more plucking till four o'clock in the afternoon, when they returned to the house. They got no respite at home after their daily labour, as they cooked the evening meal, monitored the children's progress and cajoled their easygoing husbands to shoulder a trifle more responsibility. Finally, they fell into an exhausted slumber. It was a hard life. I wondered how I could alleviate the life of the plantation ladies. The harder I thought, the more distant the solution seemed. I was at a loss for an easy way out.

Finally, the solution dawned on me while reading the works of Gandhi. Gandhi had united the diverse cauldron

resting in between a hard day's work

of Indian masses into a unified mass. The positive force of this unity proved too strong for the colonial masters to cope with. The English could not quell the force of this spiritual energy with physical violence, and India became independent. Gandhi had preached self-reliance and sufficiency to the villagers of free India in order to be truly independent. To address the cooking needs of simple villagers, he had advocated the use of biogas as a renewable, non-polluting energy supplement released from cow dung slurry. The dung was in turn used as composted manure. The smell in the dung indicated the presence of methane. On making slurry, the gas was released and trapped in a hod. A tap released the gas to the stove. It was simplicity itself that placed energy at one's fingertips. I visited the model at Gandhinagar as well as numerous others nearby. I was excited by the potential.

The first community biogas unit was installed in Makaibari with great fanfare at Kodobari village. It serviced four families. Inexplicably, the system failed within months. Excuses were aplenty. I failed to understand the cause. Why would anyone reject an easier lifestyle? Why would anyone opt for collecting firewood in pouring rain, when one had an easier, cleaner and smoke-free alternative? I mulled it over and over, but could not arrive at an answer. It was the biggest setback of my life at the time.

The plantation continued to vibrate positively in other spheres. The cow culture generated adequate compost for the tea and kitchen gardens, over 1,000 tonnes per year. The mind-boggling logistics of spreading it over 270 hectares of tea was achieved by material and verbal incentives. A handsome cash bonus for every kilogramme of compost spread, coupled with frequent reminders to each member of the plantation that healthy soil was healthy mankind, motivated the community to scale this Everest.

The village ladies had an innovative afforestation scheme. Each household raised twenty-five varieties of indigenous trees. I bought these and then turned them over to the men for

the indispensable bamboo

planting on designated areas in return for cash. The forests were preserved, and more importantly, a direct connection between the householders and their groves was established. I imposed only one condition – among the twenty-five species of trees, there should be a sapling each of a bamboo and a fruit.

The bamboo plays a vital role. Tea nurseries are raised under bamboo frames. Babies are carried in bamboo cribs. Livestock are housed in bamboo sheds and pens. Picked tea is carried in bamboo baskets. In accordance with religious beliefs, upon death, one is cremated in a bamboo pyre or buried in a bamboo coffin. Prayer flags, colourfully scripted with mantras, are strung on bamboo poles. The more they flutter, the greater the dispersion of the mantras to all corners, ensuring peace for the departed soul. From birth to death and beyond, the bamboo is indispensable. The fruit trees not only invite the birds, but are also a source of amusement and diet supplement for all ages.

As the years passed, the tea nurseries continued to bustle with activity, as people joined to fill vacancies. I was optimistic about the surge of positive development. In 1978, climatic conditions seemed ideal. The newly emergent 'First Flush' tea shoots had never been seen in such abundance. It seemed that 1978 would be a record-cropping year, with wonderful outcomes for everyone. That was, until the third week of March.

As I mounted my horse after my usual plantation round, my mood too was upbeat. As I rode on, the sky became overcast and a devastating hailstorm stripped every tree of its greenery

the First Flush

and branches within minutes. People ran for shelter, as tennis ball-sized hailstones rained down on them mercilessly. I was pounded by the onslaught. My horse, Invitation, ploughed through the foot-deep slippery ice and brought me home. Nature is a great leveller. Overnight, dreams were shattered. The plantation was desolate; all hopes and desires dashed. There would be no picking for months. All of the previous three years' work – the newly planted tea and nurseries – was lost. Fortunately, there was no loss of life – a miracle indeed. Battered and bruised, like the plantation, I too healed slowly. With the healing, it dawned on me that where there is life, there is hope, and this realisation enabled me to motivate everyone for the re-building.

It would take time to recover, and an educational programme was initiated for the healing process. The grasses, though shredded by the hailstorm, were usable as fodder. The workers commenced using them, and continued to do so, even when other vegetation re-grew two years

later. The mulching of tea had suffered for years, as I had tried unsuccessfully to emphasise to the workers that the 'community cow' – the tea plantation – should have priority. Finally, launching an awareness programme that highlighted the benefits of mulching, slowly turned the scales.

Mulching, as indicated earlier, entails placing loppings of vegetal matter like a blanket on the ground between tea bushes. The mulch takes the impact of the rain, preventing soil erosion. It stymies weed growth, but does not kill it. Thus the ground flora is preserved, while reducing weeding costs. During drought, the ground cover prevents evaporation of soil moisture,

mulching for soil protection

enabling the tea to cope excellently under this adversity. Last but not least, decomposition converts the mulch into soil that is high in humus and rich in fertility. Mulching is a simple act, but a miracle of soil rejuvenation.

Regular workshops on sustainable practices of farming were held at different villages of Makaibari. Many students from various universities from all over the globe spent time for their field dissertation, volunteering to imbue first hand, the nuances of sustainability and ethics from the ground up. Buyers visiting Makaibari invariably joined in, in case it coincided with their visit. This had a tremendous impact on their psyche. Witnessing the hard work of composting and manuring the tea slopes manually, hand weeding and mulching, planting tree saplings on steep fallow land etc. inculcated an altered sense of participation, and enhanced their ability to

sell tea lucratively. All the major overseas buyers have been with Makaibari for over twenty years. The relationship has become sustainable.

The major benefit of these on-site workshops was that they raised the consciousness of the villagers, students and buyers as holistically sustainable practices were agriculturally implemented. The other benefit of these meetings was the bonding and the camaraderie fostered between the seven villages of Makaibari.

The garden began to bloom again. But then tea prices tumbled globally in the early eighties. Out of the seventy-two plantations in Darjeeling, twenty-five were abandoned as the remaining struggled for survival. P.N., who had until then been actively leading the life of a robust planter,

the blooming garden

failed to cope with the rapidly changing needs of the hour. He thrust the entire responsibility of negotiating these troubled times upon me, and retired to Kolkata. Hard times had come. Money was scarce. It was an extremely bleak prospect to confront the unproductive four month long cold weather period. Borrowing at prohibitive interest rates seemed the only solution.

I decided to visit S.K.'s folly for inspiration, and to remind myself that nothing was impossible. My spirits soared as I gazed at the tea. Hearing a rustle, I looked up. Lo and behold, a gigantic king cobra, swaying gently with its hood unfurled, was engaged in deadly combat with a mongoose. It was a spectacular and transfixing sight. The mongoose spotted me and dived downhill into the tea. The king cobra swivelled slowly, looked benignly at me, and began sidling uphill. It was four metres or more in length, and as thick as my biceps. I pressed myself against the mountainside in order to view the magnificent serpent more closely. The cobra's size, and probably the trauma of its encounter with the mongoose as well, were enough to bring it crashing down on my head from the slope above. I screamed and leapt away in fright. I ran and ran till I collapsed on rubber legs. I looked around. The snake was nowhere in sight.

On the third day after this incident I was surprised to receive a letter from the railway authorities. P.N. had bought a wagonload of coal twenty years ago. The railway had lost it in transit and offered compensation. The money we received was just enough to see us through till the arrival of the lucrative First Flush crop. I wept at the miracle.

Fifteen years after I had tried unsuccessfully to introduce biogas plants for the plantation ladies at Makaibari, I revisited the idea in the spring of 1988 – albeit with a difference. I was riding home one day. The sun had set, and ahead of me, I saw a worker bent double under a huge load of grass. The road was steep, so his forehead almost brushed the incline as he toiled uphill. I thought him a real champion. He had been weeding in the field all day, and here he was, putting in a monumental effort for his cows. I marvelled at his feat. I dismounted, and we

huddled together for the following conversation.

I asked, 'Haven't you worked enough for the day?'

The worker replied, 'Of course, I have put in my sincere eight hours for you.'

'Aren't you tired?'

'Not really. I'm used to this.'

'How many hours do you spend on your cows everyday?'

'Two in the morning to feed, groom and milk, and two now.'

'You are amazing. Do you take a Sunday off?'

Smiling, the worker replied, 'No such luck.'

I then inquired, 'Do you realise that if you were to work the four extra hours in the garden overtime, weeding or plucking, your income would be treble your milk earnings? Moreover, you would have leisurely Sundays and holidays to boot. Why don't you go for the easier option?'

The worker replied, 'You are right. The tea fields and plantation, however, are yours while the cow is mine.'

a worker carrying grass uphill

an individual biogas unit

This was a thunderbolt. The missing piece in the puzzle of my failed biogas experiment's jigsaw slotted home. It was simplicity itself. Individual units, and not community units, needed to be installed. The very next day I re-initiated the introduction of biogas units – individual

ones – at Makaibari. It was a grand success, paving the way for the many others that followed. Miracles are simple solutions.

This success unleashed a chain of positive effects. The forests were no longer threatened for essential fuel. Care and maintenance of the cows as well as the biogas unit generated employment. The sale of milk and compost created self-respecting, grassroots entrepreneurs. The woodlands and villagers prospered.

miracles are simple solutions

Visiting buyers in London, I got a chance to meet Kiran Tawadey, the owner of the Hamstead Tea and Coffee Company, London. Although I have known her family for more than thirty years, it was a chance meeting at the Tea Board's offices at London in 1988 that cemented

our collaboration. She desired to deal in tea, particularly Darjeeling's, as it is the world's best. We had an instant empathy and over the next few years, I shared my entire knowledge about tea, with her. Today, the Hamstead Tea Company is one of Makaibari's biggest buyers and is a leading business house in distributing the finest organic produce throughout Europe.

On learning of Makaibari's sustainable practices, one day Kiran asked me, 'Why don't you get organically certified? It's big bucks'. I replied that I was not interested in 'big bucks'. She smiled in understanding at this.

Later, she took me to an organic convention at Budapest, Hungary, and then to Dornach, Switzerland. Her intention was to expose me to the worldwide organic movement, which was in its infancy then. The visit to Dornach was a pilgrimage to the global centre of sustainability. Rudolf Steiner's movement is headquartered at the Goetheanum, Dornach. As a sequel to the organic convention, she desired to expose me to the holistic philosophy of biodynamic practices as established by Steiner, whose work I had first been introduced to during my university days. Visiting the headquarters was an introduction to the finest people who propitiated these sustainable practices holistically, for Makaibari's dynamic improvement and ultimately to work and network with numerous biodynamic initiatives globally, for a better world.

Returning to London, on my way back home, I headed straight for the Steiner bookshop there. I read Steiner for the next three days and nights. It was a revelation. Steiner was many light years ahead of mankind's current evolution. Steiner revealed that the mystery of manuring could not be understood by natural science. Only by reaching into our inner-self with our thoughts, could one have access to the best methods of manuring. Steiner placed agriculture in the complete context of human evolution up to the current civilisation. According to him, cosmic rhythms have overlapped the emergence of various civilisations from the Atlantean and the Vedic Indian to the Persian. It was during the Persian civilisation, founded by Zarathustra, that agriculture arrived on the human scene. It was refined by the Egyptians and Greco-

Romans, and attained its epitome during the post-Christian era. Then, it declined to its present status in a world where humans have become predominantly materialistic. Each time the Sun passes through a Zodiacal sign – once in 2,160 years – a civilisation emerges, with its attendant agricultural evolution. The whole of modern civilisation has opted for fertilisers. This technology involves the use of chemical fertilisers that absorb nitrogen from the air. This nitrogen salt fertilises the plant, but does not enliven the earth. This empowers the farmer with false egotism, as he possesses the tools to win against nature. This is the germinating point for all anti-social forces, leading to the general decadence and the ultimate collapse of a civilisation.

The idea of biodynamics was initially conceived, when a group of farmers approached Steiner for an alternative to conventional agriculture following World War I. They were alarmed about the increasing degeneration in seed strains and cultivated plants. Could he offer a solution to improve the quality of seed and nutrition? The old seed strains of the farmers, traditionally used, were ineffective and new strains had to be brought in. This was a major cause for concern. A second group approached Steiner at the same time in 1922–23, highly concerned with the sharp increase in animal diseases, coupled with problems of sterility. They were renowned vets, doctors and owners of a pharmaceutical company. A third group of aristocrats, whose chief discipline was land and people management, approached Steiner at the same juncture with specific plant diseases that were increasingly rampant, which threatened their lifestyles in the future.

in harmony with tea

After repeated requests, Steiner finally held a historical discourse at Count Keyserlingk's palace at Koberwitz in Poland. In this weeklong agricultural course, Steiner set forth his basic view on agriculture that a farm as a whole is an organism, and therefore should have a closed

self-nourishing system. He pointed out how the health of the soil, plants and animals depend upon bringing nature into connection with cosmic, creative and shaping forces. The method of treating soil, manure, compost, and of making the biodynamic preparations, was intended to reanimate the natural forces, which were on the wane.

Steiner's works moved me to the core of my being. Where do we come from? What are we doing on Earth? Where do we go from here? These were among the most important questions in my life. I had a key to answering them at last. Returning home with a bag full of Steiner's books, I devoured them day and night for a year. Finally, the patterns began to emerge.

the spider – a friend and indicator of organic farming

I was in a perpetual state of excitement, as I continued to read Steiner while attending to the daily needs of the plantation. I took the decision to pursue an organic certification for Makaibari as it would only serve as an effective marketing tool for value addition. It would be an excellent organic launch pad to showcase Makaibari's pioneering sustainable practices in the tea manufacturing world. A few months later, Kiran called from London and asked me if I would entertain an organic inspector and a seasoned tea hand at Makaibari. A month later, Joe Smillie (Big Joe), the foremost organic farm inspector globally, along with George Neale, a leading tea broker at Wilson Smithett, London, and one of the most respected tea hands in Europe, arrived at Makaibari. George had been engaged by London Herbs and Spices as they were in search of a high class organic tea. While Big Joe tramped every nook and cranny of Makaibari, George tasted innumerable batches of Makaibari teas, in the tasting room.

Makaibari fulfilled both their stringent standards, for organics as well as quality and the certification arrived in due course, courtesy the Soil Association, UK. Unfortunately, George's sponsor, The London Herbs and Spices sourced their teas elsewhere, and George being the true gentleman that he is, saved the blushes all around by placing the entire crop of Makaibari, at one of the largest tea distributors in Europe – OTG of Germany.

Big Joe was enthralled at the sustainable practices existent at Makaibari. He was awestruck when he realised that two acres of sub-tropical virgin rain forests encircled every

sub-tropical virgin rain forests

acre of tea. Never before had he seen such permacultural practices, by the tiering of the leguminous shade trees, the mulch banks of repellent grasses, temporary legumes, fruits, and a plethora of herbs and weeds. He lauded the compost-making of the Makaibari villagers for manuring the fields, as well as the effective use of biogas digesters. The icing on the cake was the use of traditional botanicals, found in the Makaibari forests in lieu of pesticides. He was amazed to learn that these rare herbs had been in use for more than five hundred years in the Himalayas for crop protection. He was completely satisfied by our sustainable practices, which had been in place awhile.

George, a tea taster of Assam, Nilgiri, African and Sri Lankan teas, did appreciate Darjeeling Teas. He also knew the intricacies of Darjeeling Teas and the variations of the vintages. However, he was totally unprepared for the specific character and personality of Makaibari teas. He immediately drew a parallel to the legendary single malts of Scotland, and impressed on me, the importance of marketing Makaibari, henceforth as a single estate organic tea. We immediately implemented his wisdom to practise. When George placed the entire crop of Makaibari with OTG, it was a first in tea marketing in the conservative global tea trade. No single estate had ever been promoted before. Traditionally, tea had been auctioned by brokers, Kolkata being the largest centre in this regard.

Our success in selling organic tea soon led others to follow suit, thereby ushering in an attitude change. Die-hard conventional growers began converting to sustainable practices. I was no more in isolation. On the contrary, I was astonishingly propelled from laughing stock to a pioneer.

Biodynamics added a fresh impulse to work at Makaibari. Herb growing was integrated into the making of preparations or active ferments from specific herbs or plants akin to making a

alternative resources – herb growing

homeopathic dose. Steiner had created nine preparations from herbs. He had numbered them from BD 500 to BD 508. This was an extension of homeopathic preparations, which number 499.

Most of the preparations are made from herbs, encased in specific animal parts, mostly of the cow. These are then buried and lifted from the ground, according to the seasons, to create active ferments.

BD 500 and BD 501 are used for basic sprays. BD 500 is fresh cow dung encased in cow horn and buried in October, with the waning sun. It's lifted out of the ground next spring. It's ready to be sprayed as a thick spray on the grounds of tea fields after being stirred for an hour with the setting sun. This enlivens the soil forces or dark forces. BD 501 is crushed quartz, encased in cow horn and buried in spring. It's ready for use in October, when it's lifted from the ground. It is sprayed as a light, thin spray, after being stirred in water for an hour, on the same plots that have been sprayed with BD 500 the previous evening. This enlivens the light forces.

herb preparations

The purpose of the sprays is to create a perfect balance between the soil and the environment – to harmonise the two elements that would produce the right food or beverage for consumption. We harvest tea leaves that remove a lot of nutrients from the soil and plants. To supplement it, one has to manure the fields with mature compost.

Addition of the preparations BD 502 to 508, prior to application, converts the compost heap to a living organism. Yarrow flowers are dried and filled in stag bladders, and are then buried and lifted in specific seasons. Thus the activated ferment corresponds to the secretary organs – symbolic of the bladder. Minerals as well are supplemented. Similarly, encasing chamomile flowers in cow intestines creates a preparation that represents the digestive tract of the compost, when added in miniscule quantities to it. A nettle preparation regulates the temperature of the heap, while an oak bark preparation made in cow skull becomes the brain of the compost. Dandelion flowers encased in mesentery fulfil

the breathing and circulatory functions of the compost. The cumulative effect of these seven preparations converts the compost into a dynamic organism, which eats, secretes, breathes and thinks like a human body. Application of this living manure imparts a personality to the farm that is vibrant and ever evolving. Just as individuals have specific personalities, every biodynamic farm evolves as its own individual organism.

After having been organic for five years, Makaibari was finally accorded the Demeter certificate – the trademark for products of certified biodynamic production. Makaibari was the first tea plantation in the world to achieve this rare honour.

the bamboo – from birth to death

SETTING AN EXAMPLE

Peace and harmony radiated from all corners of Makaibari, so much so that even at the height of the separatist movement for Gorkhaland, when the entire district was in chaos and all the gardens were closed, Makaibari was tranquil. In spite of the fact that military forces were deployed to suppress the legitimate rights of people, resulting in much violence and bloodshed, neither the militants nor the military troubled us at Makaibari. Such are the miraculous powers of biodynamics.

When the district finally won autonomy and normalcy returned in its wake, I was pleasantly surprised to be invited by other state governments to buy land, and to practise and teach the nuances of biodynamic sustainability. They all had one desire, 'Please make our region prosper with peace and harmony'. Though flattered, I declined all overtures, saying, 'We are not interested in acquiring land. We would much rather show you how we do it.' That is how the idea for the first project was born. The project site was at Sang-Martam, a community of villages in Sikkim, bordering Tibet. Here, within the forests, tea would become a part of its agricultural diversity that included herbs, cereals, vegetables, fruits and flowers. We worked

the spring of life

the project site at Sang-Martam

with a cluster of villages in the valley to commence sustainable agricultural practices there. Cradled amidst virgin forest, the terraced fields of paddy offered some of the best views on Earth. The villages had added tea, herbs and fruit plants to the fields, and had begun horticulture, all following organic practices. We provided all the support in the form of planting material, a full-time extension officer and full technical assistance. Biogas digesters, composting and inspection for organic certification were also provided by us. We had further plans to set up a factory for product processing at the Sang-Martam project site.

Biodynamics imbued Makaibari with steady and dynamic growth. Its impact was observable at all levels of life. The forests turned vibrant with lush growth; the diversity of species increased; insects and butterflies dazzled the beholder with their calls and colours; the birds with their songs; and there were the occasional panics with panther sightings – the panther being the biggest mammalian predator in the region.

a beautiful forest marauder

Makaibari attracted professionals and environmentalists from near and afar. The greatest accolade was this eulogy from the foremost 'tiger man' in the world, Tariq Aziz, after his short but intense visit to Makaibari.

pollinating beauty

WWF Tiger Conservation Programme
WWF-India
172-B Lodhi Estate
WWF New Delhi 110 003

Date: 10 November, 1997

Shri Swaraj Kumar Banerjee
Makaibari Tea Estates
Kurseong P.O. (Dt. Darjeeling)
West Bengal 734 203

Dear Mr Banerjee,
Greetings from WWF Tiger Conservation Programme.

Makaibari was indeed a discovery for me. All the assistance and hospitality extended by

you and your family helped me to execute the objectives of my visit successfully.

The report that I will make after some time will also be made available to you. Meanwhile, I am getting in touch with the WWF-UK regarding Makaibari. I shall keep you informed of all the developments at our end.

Meanwhile, I discussed Makaibari with a friend who is also head of the World Pheasant Association in India, Dr Rahul Kaul. He will be visiting Sikkim with a team of biologists in April. On his way up, he has promised to spend a few days at Makaibari and look at the forests there. I will keep you informed regarding his plans as they crystallise. I personally feel that this group can give you a consolidated report on various faunal aspects of Makaibari forests.

I also discussed the Makaibari woods with a friend at the Centre for Wildlife Biology and Ornithology, AMU, it may be possible for the Centre to depute a few MSc Wildlife students for their dissertation work at Makaibari. The understanding is that while they learn, Makaibari will benefit in terms of documentation.

More importantly, discussions with WWF-India people suggests that Makaibari could well be chosen as 'Gift to the Earth' under the worldwide Living Planet Campaign of WWF. But prior to this, I intend to make a write up on Makaibari and have it published. Then things can take off from there – provided Makaibari contains the required components. Although, I know very little of Makaibari, I feel that it could qualify as a distinguished track of land that incorporates the three basic elements of development – biodynamic permacultural, socio-economic growth, sustainable use of natural resources and environment protection.

Please convey my regards to your mother and Mrs Banerjee.

Tariq Aziz

vibrant sights of Makaibari

Students from various universities streamed in to do dissertations, analysing the Makaibari phenomena. Most of them wanted to settle here permanently, but my gentle rejoinder to them was, 'Go home and create your own Makaibari'.

gift to the earth

AN EVENTFUL YEAR

The advent of spring ushers in the First Flush. The tea bushes are lulled into hibernation with the advent of cold weather in late November. With increasing day lengths and temperatures, the first shoot emerges in March. This becomes the eagerly sought after Darjeeling vintage, known as the 'First Flush' or 'First Pick', by connoisseurs worldwide. This is a light liquoring tea with a peach like flavour and a greenish infusion.

Throughout North India, Holi – the festival of colours – is observed to celebrate the season. The festival also symbolises love, happiness and harmony. Legend has it that an ancient and powerful king, Hiranyakasipu, drunk with power, forced his subjects to worship him. However, his son, the young Prince Prahlada, staunchly opposed him, saying that only the God Vishnu was supreme. Furious at being disobeyed, the King plotted with his sister Holika to kill his own son. The Prince's bedroom was set

Makaibari in November

alight while he slept. Oddly, it was his aunt Holika who was reduced to ashes; protected by the God Vishnu, the Prince emerged from the fire unscathed. Playing with bright colours on Holi is, hence, also a celebration of the victory of good over evil.

In 2002, the First Flush held the promise of a bumper crop. Makaibari is traditionally drought prone, but that year the early spring rains were harbingers of great hope and the First Flush commenced brilliantly. The Makaibari garden was in full bloom. In my thirty-two years of working at Makaibari, I had never witnessed such a wonderful season. That year, a special type of tea – called FTGOFPIS CHINA – was also being produced at Makaibari. Everyone was full of cheer, and celebrated Holi with great fervour.

Alas, joy turned to sadness as on 11 April, a hailstorm of gigantic proportions reduced the entire Flush to ashes. In minutes, all the tea bushes were turned into sorry looking broomsticks. In addition, the tennis ball-sized hailstones caused extensive damage to workers' houses as well as the tea factory. Mercifully, no lives were lost. Losses, however, were estimated at around one million dollars. Though under a cloud, all the group leaders began the task of inspiring and motivating the workers for immediate rehabilitation. Cow pat manure, i.e., cow dung slurried in a pit, were all dug up and around a hundred sprayers were deployed to spray the damaged tea bushes carefully. This healing spray would aid the recovery of the tea bushes. All the tea nurseries were beyond salvage – as the saplings were completely destroyed. This meant a setback of two years of hard work, as all the one year old plants in the field as well as the nurseries, were totalled. The tin roofs of many houses were destroyed by the gigantic hailstones, and all had to be repaired as swiftly and as logistically as possible. The crop loss of the First Flush, was incalculable – the biggest setback. We were headed for a bumper crop when the catastrophe occurred. The setback on annual production would be thirty per cent and losses in revenue would be forty per cent. The rehabilitation costs would add another twenty per cent to losses. This seriously jeopardised the cash flow for the next four years. To add insult to injury, borrowings from banks to tide over the calamity would inculcate another interest

Makaibari in April

burden that had to be repaid with the principal. This would take us a minimum of four years to recover, if we worked very hard – right from caring for the soil and nurturing the plants to picking, cultivating, manufacturing and, of course, selling our teas quickly and efficiently. All at Makaibari immediately went about with the business of healing in right earnest. All the buyers supported us unconditionally by buying Makaibari's at premiums. This was a precious honour, as at the most critical hour, they stood by Makaibari steadfastly. This enabled us to recover from the financial blow within the next three years.

Life at Makaibari went on. The annual inspection for the Japanese Agricultural Systems Society (JASS) was carried out in the second week of April, 2002. The Japanese government allows only JASS certified products to be marketed organically within Japan. This inspection, though a repeat of Makaibari's annual organic/Demeter (biodynamic) inspections, is essential for our Japanese collaborators – the Ishii family – who run Makaibari-Japan limited at Tokyo.

Michiko Ishii with Swaraj Kumar Banerjee

Toyota Tiger (Yoshihiri Ishii) accompanied the three inspectors from JASS as an interpreter. We were all very proud at Makaibari, when the inspectors complimented our system of recordkeeping for the hundreds of activities performed in the previous year – including soil nursing, weeding, mulching, plucking, planting, compost making, applying of biodynamic ferments, spraying of botanicals, and the making of tea and then its dispatch to various destinations all over the world.

Prior to the group's departure from Makaibari, Toyota Tiger brought us all much joy with a caring gesture of immense kindness. Kishanlal is a supervisor in the Rakti division, and he had been suffering from deafness for five years. Consequently, his work had suffered, as his hearing loss had progressed. Saddened and feeling helpless at not being able to perform his duties efficiently, he had approached me for early retirement. I had asked him to wait, as I knew that Toyota Tiger was bringing him a surprise. Before leaving for Delhi, the Tiger presented Kishanlal with a hearing aid, which he had very thoughtfully brought with him from Japan. Kishanlal was filled with joy, as he tried his new hearing aid – he could hear clearly now. He sent his special thanks to the

Srirupa Banerjee with Kishanlal

entire Ishii family for the gift, which changed his life. He remains full of energy and his work is the best from amongst all the supervisors in the field. An act of kindness that completely altered his life in turn also sweetened our entire community.

Here at Makaibari, we recently incorporated a new system of vermicomposting in order to increase the earthworm population. Earthworms are the best conditioners of the soil. The best indicator of soil health is its prevalence. The constant burrowing of an earthworm permits air to penetrate to great depths in the soil. Its food is bacteria. It devours all pathogens, anaerobic bacteria, and fungi, all detrimental to soil fertility. Hence, its abundance is essential for a healthy soil.

Vermicomposting is easily created. Earthworms are definitely the best organisms to maximise the growth of aerobic bacteria from waste. Attaining this is simplicity itself, all one has to do is create hospitable conditions and feed them with organic wastes. It is inexpensive, innovative and converts all organic wastes – including kitchen peel, into compost gold. This is the essence of vermiculture. Makaibari has copious quantities of mulch, built up over sixty years; hence introduction of vermicasts (produced by the feeding action of earthworms) has yielded fantastic dividends.

As suggested by Toyota Tiger, we decided to introduce vermicomposting at the beginning of the monsoon. In this, we benefited greatly from the wisdom of this former corporate executive, whom we here at Makaibari, have been honoured to know.

As a young lad in Japan, Toyota Tiger loved to fish. At every opportunity, irrespective of the weather and season, he went fishing. He always used the best bait to hook the fish. After a year of fishing, he learnt much of nature's wisdom and this he shared with us. He learnt that it was pointless to fish all the time. The best fish were hooked with the best earthworm bait, which he collected in summer after a good shower of rain. The following year,

as he put his newly gained knowledge into practice and fished in the summer only after the rain, he became an excellent fisherman – indeed, one of the best. He also learned that it was wiser to fish then, as the fish multiplied in that season, and so, most of his efforts yielded a good catch. He put this lesson to practice in his life, and that is one of the main reasons why he has been so successful in his career. When Toyota Tiger suggested vermicomposting at Makaibari, he had gauged that the conditions for vermicrust introduction were ideal for catalysing Makaibari's soil fertility. It certainly has proven to be true. Makaibari thanks Toyota Tiger for sharing the secret of his success with us.

This is an account of how the Goddess came to Makaibari that year.

Kurseong, is located at a distance of two kilometres from Makaibari and is the nearest town to the estate. One day, an affluent contractor living there had a vivid dream. He awakened his wife and told her that a Goddess had appeared in his dream. She had instructed him to build a temple at a specific place. He couldn't sleep for the rest of the night. At the crack of dawn, he left the house and found himself driving down Pankhabari Road (which divides the Makaibari farm into two for five kilometres). Much to his excitement, he located the spot specified by the Goddess in his dream right near the Makaibari factory.

prayer flags

At the spot, there is a huge and ancient tree which seems to have grown against all odds. It literally has a rock face and its bulbous roots have snaked at least 15 metres down into the soil. In the middle of this rocky outcrop formed by the tree, there is a large cave. Most Makaibari folk have for long worshipped the two tea bushes that have sprouted in this subterranean recess, where there is no light, water or

Makaibari tree temple

soil. It is a mystical spot, as no one could offer a logical explanation for the phenomena it contained. Where science fails, God rules, and so, the people of Makaibari worshipped there.

With support in abundance from the community, the Kurseong contractor was able to build the tree temple at the site within weeks. A number of omens preceded the formal opening of the temple. Prior to it, worshippers discovered child-like footprints clearly imprinted on the newly cemented altar, and leading to the seat where the Goddess was to be placed for worship. The very same day, a large viper also appeared at the temple. It coiled itself around the trident near it, and to everyone's amazement, slipped off by dusk. Finally, a dried bamboo grove at the entrance of the temple turned into a water source, defying logic. The water was collected in copper pots and distributed to everyone who came to watch the rituals.

On 25 May, the day of the inauguration, over 10,000 people from all over the Darjeeling region visited the site. I had been asked to release doves on the day, and a shiver passed

through my spine, as I looked at the inner sanctum. I distinctly recall seeing the face of the Goddess I worship, the Goddess Kali, and seeing it smile and bless us all. It was so real that I had goose bumps. Only time would prove whether the Goddess of Tea had truly arrived to sweeten all our lives with her terrestrial incarnation – the tea bushes of Makaibari.

An incident in September 2002 highlighted what was happening in the Darjeeling tea gardens never, or rarely, visited by their owners. The incident occurred at a garden near Makaibari, and typified the politics prevalent in most of the gardens in the region.

A Division Supervisor at the tea garden was taken seriously ill. The supervisor of a division looks after the entire management of about a third of a garden. He, or she, is responsible for the care, nurture and welfare of the tea cultivation and fertility management, as well as the welfare of the workforce in the division. Hence, he or she is an important person. So, at the request of the ailing Supervisor's wife, the Manager sent the Supervisor in his jeep to the hospital in Darjeeling for treatment.

En route, his family members decided instead to consult the most famous shaman (witch doctor), in Darjeeling. The Supervisor died during the shaman's rituals. Within the hour, the family had returned to the garden, organised their union members (over four hundred in number) and mobbed the hapless Manager. They demanded that compensation be paid immediately, since the Supervisor had not been accorded proper medical treatment. The police arrived after a few hours and rescued the harried Manager to safety. The fiasco was only resolved the next day at a meeting of the Labour Commissioner, the Union Committee and Management.

This was the situation. The Manager of a tea estate in Darjeeling had absolutely no rights. On the slightest pretext, emotions were stoked high, which resulted in chaos, pain and loss for everyone concerned. There were no winners. It served to emphasise the importance of

cumulatively practising and releasing positive sustainable impulses to avoid the occurrence of similar heartbreaking incidents. Peace and prosperity are synonymous and complimentary.

Makaibari has evolved to a different sphere of labour/management participation. The practice of holistic sustainable agricultural practices, has forged a bonding which has transcended petty politics. A daily meet is held in the factory premises with the core group management leaders as well as the village representatives. These pillars of the Makaibari community gather to discuss all problems, small and serious ones, relating to the farm, plants, forests, and the community at large. Solutions are sought, and action is initiated immediately. The motto being that 'a stitch in time saves nine'. The effect of addressing all issues on a daily basis has established a platform that has united all dynamically. Makaibari has virtually become a fertile ground for breeding peace.

Everyone at Makaibari was excited upon learning that the legendary Dr Masanobu Fukuoka, author of *The One-Straw Revolution*, would be visiting Makaibari in October 2002. Masanobu Fukuoka is a living legend. He broke away from traditional paddy farming by growing it without flooding, as traditionally practised by paddy farmers since time immemorial. He ingeniously created a farming system based on using the paddy straw as a mulch which decayed and improved soil fertility by sowing the paddy within this mulch, with a variety of other crops. The paddy was always the last to harvest, whilst the remainder were chosen selectively to harvest at convenient intervals. The waste from all the plants was recycled on the spot as mulch. This system of non-farming, as he termed it, is globally known as 'the one-straw revolution'. I admired his innovative methods of mulching with intercrops, as it was simple, effective and upgraded the soil and the environment.

On a personal note, it was a dream come true for me, as all my earlier attempts to meet him at his farm at Shikoku during my previous visit to Japan had failed owing to his frail

Dr Masanobu Fukuoka with Rajah Banerjee

Dr Masanobu Fukuoka pelleting at Makaibari

health. I was overwhelmed when I finally met him face-to-face at the Bagdogra Airport. Accompanying him was his devoted assistant, Ms Yuko Honma. Although his ninety years had taken their toll on his body, Dr Fukuoka's smile revealed strong white teeth and his laughing, tranquil eyes lit up the place. It was a special moment indeed.

Upon his arrival at Makaibari, the Doctor was welcomed to the tea factory in the traditional way with the sacred white silk scarf known as the *khada*, and the priests at the factory temple offered special prayers to God Ganesh (symbolising wisdom). The next day, the Doctor toured the tea farm at Makaibari and took a close look at all our farming practices.

Masanobu Fukuoka has devised a unique way to afforest fallow wastelands. Regions which have been devastated can be easily afforested by this method. Masanobu wanted the Makaibari community to disseminate the idea to others in the region in order to reclaim the innumerable landslide affected areas that dot the district of Darjeeling. Ms Yuko Honma taught the community how to make clay pellets. Each marble-sized pellet contains the seed of an indigenous tree. This seed remains inert until it is moistened by rain, when it germinates and sprouts. The

clay can sustain a rooted sapling even under drought conditions. We had decided to cast at least 100,000 of these clay pellets containing multiple species of indigenous legumes in the coming winter. We were also able to influence three separate paddy growers in different parts of the country to raise their crop in the Fukuoka way on a trial basis. It was a tough assignment, as I did not know how to cultivate rice and other crops using the Fukuoka way in Indian conditions; particularly, since paddy has been grown by means of flooding since times immemorial, making it imperative that the trial at Makaibari succeed and offer a positive example at the outset. I found his simple ways of farming appealing for one primary reason. Masanobu has taken the concept of mulching to a higher firmament. Unfortunately, all the trials at Makaibari using clay pellets have failed. This proved to be a major stumbling block to motivate others to follow this method at this end. The other trials with paddy growers too withered away. I still mull over the brilliance of the concept and how Masanobu has made it work at his farm in Shikoku, Japan.

Ms Yuko Honma teaching workers to make clay pellets

It was all too soon before Dr Fukuoka and Ms Honma had to depart. Their visit had been an inspiration to the Makaibari community and to me personally. His calm presence, accompanied by a disarming smile, had invigorated the entire Makaibari community. His mischievous comments were parables that were priceless. His most telling quote being – 'the ultimate goal of farming is not the growing of crops, but the cultivation and growing of human beings'.

In 2003, the Kurseong Sub-divisional Hospital became the first government hospital in the entire state of West Bengal to own a Phaco machine – the latest in cataract operation technology. This owed to the generosity of Dr Shin Yoneya of Saitama Medical Institute in Japan – a fervent supporter of our holistic work at Makaibari. It had only been a few months since I had first met him at Saitama. What struck me even then was the peace and happiness he radiated, in spite of the pressures he coped with in his occupation. I was instantly reminded of the parable my grandmother had often narrated to me so many years ago – the taller the tree, the more it bends with the breeze. The Doctor was a colossus in his profession, yet so humble.

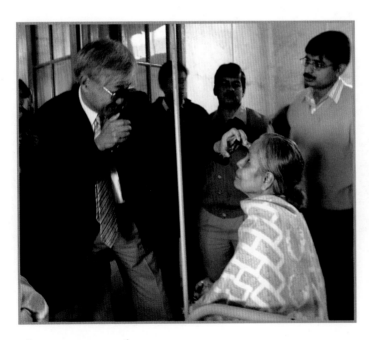

the Doctor's gift

Dr Yoneya drinks ten cups of the finest Makaibari tea daily. In appreciation, he had wanted to compliment us in his own unique way, so he gifted the Phaco machine and an Electron Microscope to benefit the Makaibari community. Realising that the machines would be put to much better use at the Kurseong Hospital, where they would benefit all the poor of the region, I began the protracted process

of fulfilling all the bureaucratic norms necessary to import the Doctor's gifts and place them at the hospital in adherence to the letter of the law. This process was slow and arduous, but finally thirteen months later, the machines arrived at Makaibari.

Dr Yoneya is not only a man with a heart as big as a barn door but also one of the finest eye surgeons in the world. When the machines arrived, the Doctor also conducted free eye camps for the community between 27 and 31 December. This positive act had multiple benefits; not least of all for the sixteen poor people who had been in darkness until then but were now in a radiant paradise, having had successful eye surgeries with no complications. It also earned him the enormous gratitude of the Makaibari community and all of Darjeeling.

tennis ball-sized hail rained mercilessly

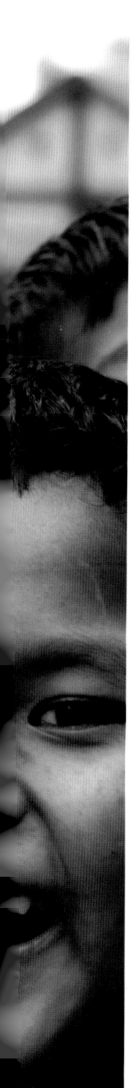

THE COMMUNITY

The story of Makaibari would be incomplete without mention of the village children. Their education is state controlled, but the plantation contributes the land and a part of the building expenses. Though government employees, the teachers are influenced inevitably by the forces of biodynamics that surround them at Makaibari. Interest in the work permeates, resulting in an appreciation of natural rhythms and the positive role of humans within it in their discourses. It is a slow and subtle phenomenon.

Steiner's teachings are beautiful. The light from the cosmos creates life. The fusion of the two results in love – the most powerful force in the universe. Force can evince quick results but these are not sustainable, while love and patience create awareness on a long term and sustainable basis. One outstanding contributor to creating such awareness was a visitor from Germany – Frauke Mai. She organised a plastic litter collection through innovative songs and dances with the children's playgroups. The children took to it like a duck takes to water. Every Sunday, they bring in the litter and receive a small reward. They also participate in a group activity to learn about cows and compost making in a fun way.

educative laughter

health and hygiene class for children

The ladies action group followed the example of the children and began participating in similar activities. They watched interesting documentaries on childcare and family planning. Motivated by their newly gained awareness of these issues, they devised a novel scheme that gave incentives to limit the family size to two. This scheme entailed granting extended maternity leave as well as handsome cash awards for sterilisation. Free contraceptives were also distributed by the tea garden's doctor, paramedic and nurse. The scheme was a huge success, resulting in a negative birth rate. Makaibari was the only tea plantation in India to hold this unique record. More importantly, it meant that the children are properly cared for, and they can look confidently into the future with hope. The community ladies are responsible for this feat.

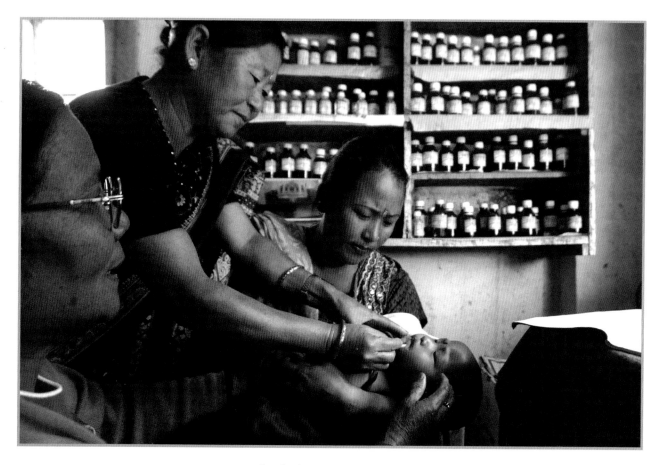

the ladies action group

Initiation of the vasectomy programme, on the other hand was plagued with obstacles. Excuses were many. Some of the most popular included: 'We will become weak.' 'We will be impotent.' 'We cannot make love any more.' It was difficult to overcome these stumbling blocks. Eventually, I was able to persuade three men, all with a brood of children, to volunteer for the procedure in exchange for favours they had sought from me. Off they went to the hospital with fanfare. The elderly nurse, who had seen it all, curtly ordered them to change into hospital smocks. Two waited, as their friend lay in a curtained area of the room. Watching the doctor arrange his tools, the volunteer moaned pitiably, 'Oh God, have mercy on me.' That was enough to send the other two scampering through the first floor windows to the busy street below. Pandemonium broke loose, as traffic came to a standstill. The bewildered runaways in their hospital smocks ran crazily through the traffic for their lives. The commotion distracted the doctor and nurse, and the third volunteer seized this opportunity to escape in the opposite direction. The trio returned after a week to the plantation with wild tales of their escapade.

This ensured the men would not participate in the programme. For my part, I realised that guile was no substitute for awareness. The men would join the programme only when they were ready.

This failure of the vasectomy programme was offset not only by the success of the ladies action group but also the entrepreneurship exhibited by the milkmen and vegetable growers on the plantation. These workers sold their surplus in the local market. Word soon spread about the high quality of their products, increasing the demand for them. Even though the workers added a premium to their price, the queues only kept increasing. All Makaibari produce at the local market was quickly sold out. This was the best substantiation of the biodynamic spirit of Makaibari.

The Makaibari Joint Body (MBJB) comprises of elected members from the Makaibari tea farm's seven villages. The MBJB is an elected committee of the seven Makaibari villages, namely, Kodobari, Thapathully, Foolbari, Cheptey, Koilapani, Makaibari and Halder Kothi. Elections are held every three years. There are twelve ladies and six men in the committee. I am the only permanent member. This was the committee's decision, chosen to exercise my special rights in the event of a deadlock on an agenda. It's been in existence since 1991, and to date, I have not been called in to cast my vote.

The committee is responsible for all social development criteria, not only in Makaibari but in the neighbouring villages surrounding Makaibari. A small premium is set aside from all sales, which has been the seed corpus for funding the initiatives. These are predominantly micro-loans for small start-up businesses, extension services for animal husbandry, building water-borne toilets, training of paramedics for basic health and hygiene, specialised medical care, pre and post natal care, afforestation projects with World Wildlife Foundation and homestay constructions. All these and more have been successfully implemented by the MBJB.

What is remarkable is the role played by ladies. None of the ladies are bankers, economists or stock brokers. They are all high school graduates. However, when it comes to deployment of funds for community upliftment, they instinctively home in on the most suitable projects for their locale, which ensures its success. Unless ladies are empowered, no community can march forward positively. A majority of the successful schemes deployed at Makaibari, have all been managed by ladies.

Community development is one of its primary goals. The funds of the MBJB are premiums received from tea sales in Europe and USA via the Fairtrade Labelling Organisation (FLO). The positive nature of the work done by the MBJB is best illustrated by this simple story about Jamuni, whose life was transformed a few years ago when she received a cow as a gift from the MBJB. The cow was donated to Jamuni as part of a programme operated by the MBJB that entails working with families and communities who want to raise livestock and gain farming skills.

the Makaibari Joint Body

Jamuni Mangarni is a farmer who grows tea, corn, fruits and beans. She lives in Thapathully village in Makaibari, where the people call her Jethi. For years Jethi worked hard. It was a constant struggle for her to survive and make both ends meet. Her husband provided little support, he had moved to Kolkata and rarely came home. Jethi's life was one of daily toil. When she received the cow as a gift from the MBJB, it changed her life forever. Not every family in her village had received a cow. One such family offered Jethi help to feed her cow in exchange for its firstborn female calf. Jethi cared for the calf when it was born, and a few months later, turned it over to the family that had helped her. Others in the community who helped Jethi build a cowshed received the second calf. This way, in time all villagers benefited from just one cow.

Jethi's life improved after the first calf was born. Her children drank milk and became healthier. She sold the extra milk and improved her lifestyle. The manure from the cow fertilised her fields, improving the quality of her crop and yields. She now had surplus grain, fruit and vegetables to increase her income, and further augmented it by selling excess manure to the Makaibari tea farm. The MBJB then gifted her a portable biogas and cooker which further improved the quality of her life. Now she had a non-polluting, renewable source of energy on tap for fuel. The woods were no longer threatened and were able to prosper too.

the mooing empowerment of Makaibari ladies – Jethi

Jethi's success was due to the way she maintained her cow. She used a system called zero grazing. This means she penned her cow, fed it cut grass and did not allow it to graze away. This allowed her to watch her cow and instantly remove ticks and parasites. The cow was healthier and no vet was needed.

Her success spread, and in this way, all the villages in and around Makaibari benefited. Her husband too returned home. Jethi's hard work had paid off, and the simple cow had played an important role in not only bringing her family together, but also turned an impoverished community into a self-respecting grassroot entrepreneurial one. Recently, Jethi gifted the MBJB 10,000 rupees to enable the organisation to initiate another 'Jamuni Chapter' in the region.

plastic litter collection by children

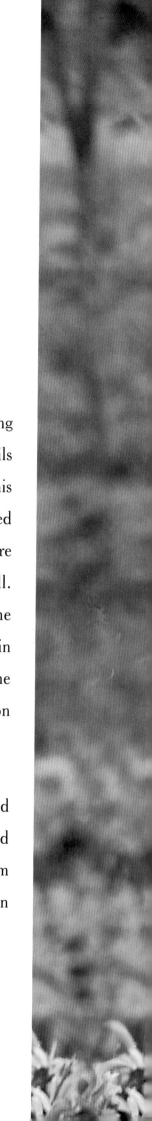

MAKAIBARI TEA

Over the years, I learnt the secrets of making Darjeeling vintages. The orthodox way of withering entails exposing the freshly plucked leaf to an air stream. This eliminates the excess moisture inherent in the leaf. Then it is rolled to catalyse fermentation and to impart leaf style. The cells are ruptured, and their essences are kneaded into the roll. Fermentation is essentially a process of decay. The same phenomenon occurs when a fruit is picked and allowed to remain for a few days in the pantry. This process of death releases the enzymes that are so essential for developing the aroma and infusion of the tea.

After the initial First Flush bounty, the tea bushes are lulled to a non-productive period of four to five weeks. This is called the '*banji*' period between mid April to mid May. From November till April, there is scarcely any rainfall, so the region acquires a dry, arid look.

tea plucking

Makaibari tea leaves

During the *banji* period, fast moving rain bearing clouds sweep up from the Bay of Bengal, and on collision with the Himalayan foothills, inundate the Darjeeling district with short, but fierce bursts of precipitation. These are the awe-inspiring norwesters. In a flash, dark clouds appear out of clear blue skies. The lashing is intense, with copious rain, accompanied by streaks of lightning and deafening thunderbolts. God help those who do not unhook their electronics, as they blow up instantly. Within an hour, it is all over and the sun appears with clear skies. This is a magical moment. Overnight, the region is a riot of green and is abuzz with the emergence of all life forms. Greenflies to the tiger, all rejoice and breed profusely in this period.

a production process – withering

The swiftly multiplying greenflies, feed on the emergent Second Flush shoots that break the *banji*. The growth of the leaves is stunted due to the greenflies feeding on them. These affected tea leaves are plucked and manufactured as the rare Muscatel Second Flush. This fetches high premiums.

The Orthodox Black Darjeeling – which is globally renowned – is normally made from these two critical periods. In the orthodox manufacture, the leaves are withered after plucking for fourteen hours. Seventy per cent of the moisture is removed here. The withered leaves are next rolled. The purpose of rolling is two-fold; firstly, to impart leaf style. The whole leaf normally marked ftgfop1s, fetches the highest price as opposed to broken grades, normally tea-bagged. Great care is ensured at rolling to extract high percentages of the leaf grade. Secondly, the rolling catalyses fermentation. The fermentation is stopped by firing the teas in a drier. The time of firing the tea varies seasonally. The rare muscatel is so called, as it resembles the flavour of the vintage muscadet reds of Bordeaux. As a matter of interest, the muscatels have a bright coppery infusion as well as a deeper colour in the cup.

a production process – drying

the delicate touch – sorting

The monsoon sets in with right earnest in mid June. The withering process is difficult to manage, as humidity is at a hundred per cent. This is the ideal time to manufacture the green teas. Green teas are unfermented teas as opposed to black fermented teas. The freshly plucked tea leaves are steamed to arrest the fermentation. Then they are rolled for styling and a final run through the drier ensures that the moisture content of the tea is less than two per cent.

The master tea maker is a person who is one with the practice of tea making. He, or she, ensures that every roll is imparted an aroma with its own distinct personality. The slow and natural process of decay is accelerated by heat, light and humidity, and at a critical time of climax, the fusion is arrested. The process is a mystery; the quality is the manifestation of the magic.

the Silver Tips Imperial tasting

Indrey Sarki, is a master tea maker. He has been the fermentation expert at Makaibari for the last ten years, after training with his mentor for twenty years. His nose is like a bloodhound's, and it has been honed to detect the critical moment when the fermentation should be arrested by firing it in the drier. His devotion to his craft has been a major contribution to the specific personality of Makaibari teas.

Playing with the fermentation process results in various types of teas; green tea is steamed on picking, and thus, not fermented. This explains its natural curative potency, which is particularly retained in organic leaves, blessing the tea with a wide range of therapeutic properties. A close friend of mine was once afflicted with one of the worst kind of hepatitis.

Apparently, there was no cure. I recommended drinking ten cups of green tea daily. Amazingly, my friend recovered completely within a year. In Darjeeling, it was called a 'medical miracle'. I attribute it to the green tea.

A growing awareness about health and the environment swayed me to adopt green tea making. The demand for this magical brew is explosive. The world is a global village, and it appears that the tea for the next millennium will be green.

Makaibari has a patented green tea which is the Silver Green. In July 2007, an accolade has been awarded to this tea from USA, as the tea with the greatest content of anti-oxidants. The Bai Mu Dan is a specially crafted green tea, which retains the entire shape of the plucked two leaves and a bud. This is an extremely delicate tea and is specially ordered by mail from global connoisseurs. The Silver Tips Imperial is the current record holder as the world's most expensive tea. The tea is harvested in and around the full moon, which severely limits its production. Quality and quantity are not synonymous and the limited Silver Tips Imperial affirms this. The tea is truly a libation and is an offering to all who drink as they are our gods at Makaibari. This too is a Makaibari patent.

connoisseur's delight

THE TEA DEVA

The Tea Deva was first found in 1991 in Makaibari and baffled entomologists worldwide. It looked like an insect, but no matching insect could be found in the textbooks. What was remarkable about it was that it was the exact replica of a tea leaf – in every aspect and detail. In the summer, it carried the typical signs of a fresh, new leaf, while in the winter, it showed the natural blisters that affect tea leaves. Indeed, when in 1995 a hailstorm devastated one section of the tea plantations at Makaibari and left large patches on the tea leaves there the same injury marks were seen on the Tea Deva in a different and undisturbed section of the estate.

While scientists at the Zoological Society of India and the University of Hohenheim conducted research, I decided to call this unique life form, the Tea Deva. For me, it was a divine manifestation. It was one that had made its appearance, or perhaps reappearance, as a result of the religious diligence with which biodynamics was, and continues to be, practised at Makaibari. As Rudolf Steiner – the father of biodynamic agriculture – has stated, if all agricultural practices are truly holistic, then the principal crop will be reflected in mimicry. Clearly, we must be doing something right at Makaibari.

the Tea Deva

the replica of a tea leaf

Finally scientists from Calcutta University and the Zoological Survey of India gave their verdict, identifying the Tea Deva as a member of the Phillidae family, which is adept at mimicry. This family includes 'walking insects' that resemble sticks, twigs and deadwood, among many other forms. However, none replicating a tea leaf had been recorded until the discovery of the Tea Deva in Makaibari.

The Tea Deva is picked up ten or so times in Makaibari each year. Indeed, each time a Tea Deva is spotted and picked up by a worker, he, or she, is handsomely rewarded – especially, if one is picked up for visitors staying on the estate to see firsthand this miracle of Makaibari. This incentive has resulted in a habit of looking at nature more closely among members of the community – young and old, women and children both.

In spite of the many pairs of eyes looking for it, the Tea Deva is not easy to find. It is, after all, very difficult to spot, as its green colour and tea leaf shape act as excellent camouflage against the tea fields. For example, no one could find one for the visit of Ms Keiko Taneda (a journalist from Japan) and Mr K. Kikuchi (a distinguished photographer and expert in Chinese tea) to Makaibari in August 2001. However, a ranger was able to spot one to welcome the visit, as it were, of Stephen Lee and Thomas

the living form of a miracle

Mesher (both friends of Makaibari from the United States of America) in the autumn of 1999. One was also found for the distinguished Dr Fukuoka's visit to our community in late 2002. Each time that it is found and picked up, the Tea Deva is also released back into its tea habitat (after its habits have been studied). And each time it is released it literally vanishes in front of our eyes, so perfect is its mimicry.

the perfect mimic

The Tea Deva remains a miracle of Makaibari.

Swaraj Kumar careening around the plantations

LORE OF
THE LOGO

In 1983, my wife and I were travelling by air from Bagdogra to Ajmer to place our elder son, Udayan (then ten), at a boarding school. The school is located 1,800 kilometres away in Rajasthan in the western part of India. We first took a two-hour flight to Delhi, followed by another thirty-minute flight to Jaipur and finally a two-hour drive brought us to Mayo College. For the journey, I had picked up a magazine to read. The cover story was about a psychic and the article was titled, 'The Man Who Sold Luck'.

I was engrossed by this story about Mr Krishna, a man from Bangalore, who harmonised the vibrations of troubled people by designing a logo for them. This logo would then bring them peace, harmony and material success. Having for long dealt with the vibrations of plants, I thought that this was indeed a great soul. The vibrations of humans are far more complex than those of plants, as is harmonising them.

the tea factory

the new Makaibari logo

I wrote to Mr Krishna, and after lengthy correspondence, he stated that our 140-year-old logo was unsuitable. I consulted all the members of my family for their opinion. They were all against any change and laughed off Mr Krishna's opinion. Moreover, the psychic-designer was very expensive. As luck would have it, an insurance policy of mine matured at that point, and its value covered the fees exactly. The family had no objections if I were to pay from my own resources for the logo. The money was sent and the logo arrived a few months later.

Believe you me life has never been the same since! The meaning of the logo, however, remained a mystery, and it was not until sixteen years after its incorporation that Mr Stephen Lee interpreted it.

Mr Stephen Lee is a prominent tea and herb importer from Portland, USA. He has been associated with the tea business for more than twenty years. Recently, he bought a tea company called Teaports. Upon entering the company warehouse to take stocks, the first sight that greeted him was that of Makaibari tea chests, bearing the estate's logo. Stephen was hypnotised by the logo, and an image of it stayed with him throughout the night. The following morning he decided to take the earliest flight to Makaibari.

Stephen and his Chief Executive, Tom, arrived at Makaibari at three o'clock in the afternoon, after flying non-stop from the west coast of America. They were met at the Ganesha Temple in the factory by the community ladies, who welcomed them with traditional *khadas* or silk scarves. A conch shell was blown as a symbol of peace, and mantras were chanted by the temple priest. Stephen and Tom were then taken on a tour showing them the manufacture of conventional black tea, after which they were scheduled to meet community group leaders. As they were being introduced to the community committee, a worker came in excitedly. In her hands she was holding a Tea Deva, which she had found only moments earlier. Stephen and Tom were mesmerised by the unique insect and filmed it extensively. Both were completely amazed by this Makaibari specialty – a moving tea leaf!

the logo

As we walked through the tea plantation the next day, Stephen picked up a flower off the tea bush. It was a delicate white flower with yellow pollen in the centre. Suddenly, he stopped and asked us to have a closer look at the flower. He turned it around for us to look, and astonishingly, the pedicle of the flower had the same shape and colour of the Makaibari logo. Thinking it an accidental occurrence, all of us collected a number of flowers, but were startled to find that all of them had the same configuration and colour as the Makaibari logo. I was bowled over in surprise by this revelation.

Stephen was happy to have made the journey to Makaibari, opining that it must be the home of tea, a cup of which cheers everyone internationally. The community at Makaibari, in

turn, was intensely grateful to Stephen for affirming our impulses that Makaibari had the blessings of the cosmos. Thus began our partnership with this unique messenger.

The night following Stephen's departure, I had a dream – a very vivid one, related to an ancient Chinese folktale I had read recently. It went something like this. According to a famous Chinese legend, during the reign of Yuen Ty of the dynasty of Tsin in the third century AD, there lived an old woman who had a magical cup. I dreamt that every morning at daybreak, this frail old lady, whose name was Ming Mai, sat in a corner of the marketplace with her magical cup of tea. From dawn to dusk, people thronged to her to buy and drink the tea from her cup, which never emptied. All left satisfied, as the tea fulfilled all their hopes, dreams and desires, and their pains evaporated. The queues kept getting longer, as her popularity increased. The money Ming Mai received, she distributed to the poor and needy, keeping only a minimal amount for her daily sustenance.

In due course, the news of her popularity reached the Emperor Yuen Ty. He was envious of her fame and goodwill and sent his armed troops to confiscate Ming Mai's magical cup. When his troops failed to get it from Ming Mai, he threw her into his darkest dungeons. Late at night, as moonbeams filtered through an opening in the high ceiling, Ming Mai flew out of this aperture, and across Tibet. She scoured the Himalayas for more than fifteen hundred years, till

tea flowers with the Makaibari logo

finally in 1859, she found her ideal resting place in Darjeeling. And, this was how Makaibari came to be the first tea garden in Darjeeling.

To date, her cup continues to sweeten all who drink Makaibari tea, for one not only imbibes a high-quality Darjeeling Tea but also the spirit of Ming Mai as the spirit of Makaibari. The five petals of the Makaibari logo represent the five continents; the central circle represents the heart of Makaibari, and the ring ensures that all benefit equally when the Makaibari spirit is imbibed.

the spirit of harmony

Lore of the Logo

SPREADING THE SPIRIT OF MAKAIBARI

There are over ten thousand tea estates, employing about fifty lakh people, through the length and breadth of India – from Darjeeling, Assam, Tripura, Arunachal Pradesh and Nagaland in the east to Himachal Pradesh in the north, and Tamil Nadu, Karnataka and Kerala in the south. In 2001, one of our community members was elected from amongst all the tea plantation workers in India to attend a four-day global conference organised by the Fairtrade Labelling Organisation (FLO). Ruplal Rai, from our Koilapani, left for Lingen in Germany to attend the conference on 16 September, 2001. It was like a dream come true for Ruplal, and for all of our community, that someone from Makaibari had been given such a great honour.

On his arrival in Dusseldorf, Ruplal thought he was in paradise. He would not have believed that such wonderful infrastructure could exist in this world. He was amazed at the beautiful autobahns

the wonder of Darjeeling

the trail blazer – Ruplal Rai

(highways), and the high speeds at which traffic moved on them. The houses, the avenues, the shops and the cars – he was awed by them all. He could not believe that such order, as he found in Germany, was possible.

At the conference, he was greatly cheered to find a facsimile from me in Nepali waiting for him. He felt very proud to know that the entire Makaibari community was supporting him – he was alone in the gathering of five hundred visitors, for whom the organisers had searched in order to deliver the fax. He felt strong, and his resolve to present the spirit of Makaibari to all those gathered at the conference was very clear. He realised that in spite of all the material glitter that surrounded him, Germany did not have the inner radiance of the people living peacefully with nature that was in his own heart from his life in Makaibari.

He impressed everyone with his wonderful accounts of life in Makaibari. As he also did when he told them about how the Makaibari community had evolved to joint forest management, thus addressing Agenda 21, which all Western governments were debating on incorporating, for rural lifestyles this millennium. He was so inspiring that he was invited to address two groups in Dresden and Leipzig, as well as the German press. This simple tea farmer from Makaibari charmed everyone!

He was then invited on 25 September to meet the German President, Mr Rau. The President asked him what he should do first – drink the tea, watch the video of Makaibari, or read my book *Das Wunder von Darjeeling*, all of which Ruplal had presented him with. Ruplal replied that as Mr Rau was the German President, everyone at Makaibari believed him to be the Trinity, so he could perform all three acts at once, if he so chose. Everyone laughed at this, and Ruplal became the darling of the German President.

We welcomed him back home in India on 27 September. As he himself is one of its wonders, Ruplal has also done wonders for Makaibari.

In February, 2003, I attended BioFach in Nuremberg, where this annual organic trade fair had been held for the past three years. BioFach is the largest organic fair in the world, and exhibitors from all over the world attend it. Fourteen years ago, it had been a homely, small-scale affair with about fifty exhibitors at the most. It commenced humbly at Frankfurt Messe by a group of dedicated organic producers at Germany. Their intention was to introduce the benefits of being sustainable, as well as to create consumer awareness on the holistic impact of consuming organically grown food. One of the principal promoters was Demeter-International, based at Darmstadt, 30 kilometres from Frankfurt. I had visited them and was impressed by the dedication of the small group. Ever so slightly, it began to grow over the next few years. The concerted efforts were beginning to pay. It had become sizeable by 1996 with over a thousand worldwide exhibitors. It was the first time we had participated as an exhibitor with support from the Tea Board of India. It was a fantastic experience, as it opened vistas from non-traditional customers who were eager to launch a high quality organic Darjeeling Tea under their label. The impact was so great that the Hampstead Tea and Coffee Company completely revamped their range to partner Makaibari for this prestigious organic show. It grew to such proportions that the German government allotted a permanent site at the Nuremberg Messe in 2000. It had become a mammoth event with ten halls, each the size of a football field, devoted to it. Organics was finally mainstream!

Most of Makaibari's traditional European supporters attended this fair and over four days, I was able to meet all of them at our joint stand with Hampstead. It was equally thrilling to witness the support for organics as well as fair trade at the fair. It was obvious that huge numbers of people the world over, were becoming more aware of the critical issues surrounding sustainability. Fair trade and organics seemed to be well on their way to becoming mainstream with greater sales (and their attendant problems!). Eco agriculture was now economically viable. This was the catalytic assurance needed by farmers globally to convert to sustainable agricultural practices. Makaibari, was routinely thrust to the forefront of all tea buyers globally, seeking high quality organic tea. More importantly, an increasing number of people started visiting Makaibari to ascertain for themselves, the positive benefits of sustainability. Oratory is a poor substitute for visual impact, and all visitors are moved by the impulses released at Makaibari, that address economics, politics and the environment.

In 2003, my wife Srirupa, who is an active Director of Makaibari, as well as an active elected member of the MBJB, visited France as a guest of the Fairtrade Labelling Organisation, the French Government and Max Havelaar. The Fairtrade Labelling Organisation (FLO), based in Cologne, Germany, allocates various brands for identifying Fairtrade products in Europe, USA and other countries. In France, Scandinavia and Switzerland, it is labelled Max Havelaar. In Germany, and USA, it's Transfair, and in UK, it's

Srirupa – after a successful Joint Body project

Fairtrade. All use a common logo of the ying and yang – under the FLO umbrella. Srirupa was there to showcase the sustainable philosophy that had made Makaibari the wonder of Darjeeling, and made presentations at various forums – both private and public.

In 2003, between the end of June and the beginning of July, we went on a trip to Japan where we were hosted by the Toyota Tiger family – our wonderful Japanese Makaibari family. One of the highlights of this trip was a visit to the legendary Yamashita San's tea factory at Kyoto. Yamashita San has been a recipient of the Japanese nation's award for excellence twenty-five times. Receiving it once is a 'dream come true' for anyone! The reason why he had received this accolade so many times was made amply clear over the following hours. He is a true icon for all artists!

Gyokuro Green Tea is one of the most difficult teas to make. It is specially grown and plucked, and it is entirely handmade. Manually producing this needle-shaped tea takes over eight hours. The production process consists of five distinct, arduous manipulations, each over an hour long in duration. It is gruelling work, but it is an art form. Yamashita San's passionate devotion to mastering and perfecting its techniques for over fifty years is what makes him such a special being.

The first stage in the process takes two hours during which one needs to remove the extra moisture from the steamed leaf. As I was midway through this process, while making my first batch of Gyokuro tea, journalists from all parts of Japan began arriving in small batches. One asked me what was I feeling at that moment. Not pausing from my labour, I replied, 'like an out of tune violin in the Yamashita Gyokuro symphony'. This made headlines the following day in one of Japan's leading dailies. Over the next eight hours, Yamashita San patiently and gently taught me the nuances of making Gyokuro tea. I was greatly honoured. This was followed the next day by an in-depth discussion with Yamashita San about the details of cultivating,

pruning, plucking and shading the tea. We discussed all its aspects threadbare so as to produce the finest Gyokuro someday at Makaibari.

Another highlight of this trip was the discussion that followed a presentation I made at a famous organic restaurant in Yokohama, called Kapukapu. When we arrived at the restaurant, it was filled with people, who had travelled from far and wide to be there that day. One of the most striking things about this gathering was that about ninety per cent of those present were women, the male representation being no more than ten per cent.

Rajah at a Tokyo symposium

After I had made my presentation on Makaibari, a question and answer session followed. One of the first questions I was asked was as follows: 'What was the crop reduction in switching from conventional to organic farming practices, and how did you control pests and insects?' This was indeed a pertinent question. I replied that our recorded experience revealed that in the first year of conversion, the crop loss was thirty per cent, but it progressively stabilised at fifteen per cent of the conventional yields four years after conversion. More significantly, the price jump was forty per cent in the first year itself, so in reality, all consumers paid a better price for quality and the net result was gain. In spite of the huge crop loss in the first year, the quality jump was so noticeable that buyers paid a forty per cent premium on quality alone. The balance sheet reflected a gain right from the start. Subsequently, the publicity that accompanied our pioneering sustainable work in Makaibari was so huge that marketing it was a simple step in capitalisation, which brought in even more rewards for the community.

The most important factor in conversion is compost management. The villagers in Makaibari own over 1,000 cows, yielding about 1,200 tonnes of manure. This is sold back to the management to fertilise the tea and their own crops and vegetables. The milk from the cows – 3,000 litres per day – is sold in nearby Kurseong at a premium of thirty per cent. This is instantly sold out on arrival. Although this milk is not certified organic (as in India the concept of organics is a distant reality for the marginalised consumer), the quality of the milk is obviously organic and attracts the consumer to it. Thus, it proves the point that anyone will pay more for quality.

In addition, supplied biogas units enabled the availability of non-polluting

the soil health initiator

the best of Makaibari teas

renewable energy on tap, which relieved the pressure on the woodlands for fuel. The water which slurries the cow dung releases methane, which is then trapped in a hod and tapped as needed. Women, who traditionally bent their backs from the crack of dawn to collect firewood, were relieved from their backbreaking chore. The time saved, three hours, has accorded them considerable leisure time as well as an additional income from the sale of milk and compost. Furthermore, an enormous amount of money has been saved by the improved organic techniques for growing corn, fruit, and vegetables. The fresh and pure commodities have considerably improved the health of the Makaibari community. 'Going Organic' has many bonuses as well as fallouts. These are the hidden benefits, which most people ignore.

At Makaibari, two hectares of virgin sub-tropical rain forests surround every hectare of tea. Furthermore, the tea is grown through practices of permaculture. Tea is one of six stages of permanent vegetation, occupying the fourth tier of diversity. This ensures a food web that is

in complete harmony. All species of insects and pests exist in Makaibari. However, the ecosystem is so finely tuned that there are no alarming explosions of any pests harmful to the tea. Predators keep their population firmly under control. Thus, all life forms coexist, and natural laws control the ebb and flow. In Makaibari, the flavour in the 'balance sheet of life' is the focus rather than the flavour in the balance sheet exclusively.

In early 2004, Lafi Bala, a Fairtrade organisation based in Montpellier, France, invited me for a series of workshops. They actually desired Srirupa, but as she had to undergo a surgery, I was the surrogate. I participated proactively in over a dozen high schools in five cities in the south of France. This was preceded by a three-day seminar for about seventy European high school teachers on fair trade initiatives in Makaibari. For them, this was a window from which to glean firsthand how the principles of fair trade actually worked in practice. The teachers taught their students what they learned which could in turn increase Fairtrade sales, the motto of FLO being, 'Fair Trade, Not Aide'. Due to Srirupa's visits to France Makaibari has prominence in supermarket counters there and is well recognised. The purpose of the visit was to persuade high school children to support Fairtrade labelled products which in turn would uplift marginalised folk who produced it economically from the small premiums paid by them. I had to literally tell the teachers as well as the children, step by step, how the small premium was directly sent to the MBJB, and how the projects initiated from these funds created a grassroots self-respecting entrepreneur.

I had also been invited to travel with the Lafi Bala bus across the south of France. The words *Lafi Bala* come from the African country of Burkina Faso and mean a friendly greeting of well-being. The Lafi Bala bus was colourfully painted with the organisation's logo. It had simple knock down displays – an entire bus full of it, immaculately arranged. Lafi Bala had been promoting Fairtrade cocoa and coffee from Africa and South America for a decade. Tea had only just been introduced into their portfolio. Hence, while there was a wonderful array

of road exhibits on coffee and cocoa – each the result of years of dedicated research – at the back of the bus, alas, the tea cupboard was barren. They had little to no material on tea, and indeed, they had little idea about what to do with me after having invited me. There was no road map for tea, while the displays, games and educational activities for coffee and cocoa were bewitching. The onerous task of bearing the torch for tea was singularly mine. It was thrilling for me to innovate on the spot, with my limited display material (as I had not been informed earlier of their inadequacy on tea displays) and samples, as we hurtled from one French city to another in the bus. Fortunately, most of the schools provided overnight lodgings, and excellent food, so the strenuous rides, followed by presentations at different venues everyday, was adequately compensated. A year later, most of our French customers had a significant increase in Max Havelaar labelled sales of Makaibari teas.

When we arrived at a large school in the suburbs of Montpellier, I was simply thrown into a large classroom of students with minimal knowledge of English and an interpreter with no interpretation skills. But this is just the sort of challenge I revel in. I targeted a student who had some spoken English skills and wove a story around him and his family with a set of simple questions. Using this family as a nucleus, I extrapolated this simple agenda to the global family. I asked Thomas, my student helper, about what happened to a family that did not communicate with each other. His reply was that then unhappiness and misunderstanding resulted in needless squabbling and acrimony. I then asked, what about a family where members had time for each other? A cheerful environment, where all members were happy, confident and supportive of each other and a radiance that carried itself outside the house to others resulted, was the prompt rejoinder.

I next linked this microcosm to the macrocosm, expounding that communication is the key to releasing dynamic synergy for holistic global sustainability. Email is so cheap and effective that communication from any corner of the globe is easy. Truly, the microchip has converted the world into a global village. The reason I was there was to share and interact dynamically

with them, and to impart the benefits of living holistically. What I had to say was simple. Healthy soil is healthy mankind. To support this truth is to buy organic and Fairtrade foods; organic because it is both healthy and environment friendly, Fairtrade because it ensured a fair price to the marginalised grower from the third world whose annual income was $150. This figure shocked the entire audience and from that point it was easy to lead them in the desired direction. Most of the children present there received a monthly allowance of $150. Unanimously, they committed to be supportive of Fairtrade and organic products.

Where I had the benefit of a good interpreter, the impact was overwhelming. One such supporter was Rachida Chatt an elegant Moroccan, English teacher. Her sole passion in life was teaching English, and her command and diction were excellent. The students too held her in high esteem which was reflected in the packed audience. Her interpretations were spiced with her own innovations as was apparent from the shrieks of appreciative laughter and applause that followed a joke or a salient point on sustainability. Fortunately, she is a tea fanatic and it was easy for me to repay her with Makaibari vintages. She called me a few days later and informed me that it was the first time she had tasted something as divine as the Makaibari pure Darjeeling Organic Tea. In the past, she had only drunk the cheap and cheerful varieties promoted by multinationals. She was now a Makaibari addict forever. Moreover, to my immense delight, some of the girls in her class had also become tea lovers.

The following week, I was the main speaker at one of the forums. The discussion and interaction after my presentation were intense as the majority of the French people present there were totally oblivious of Indian culture. Moreover, most of the teachers in the audience were totally in the dark about the nuances of organic agriculture and fair trade. To most of them, organic simply meant pesticide-free hence the question and answer session that ensued after my presentation was lively as well as exciting for all of us and built many positive bridges.

Most of the audience could not understand the criteria for Fairtrade tea vis-à-vis Fairtrade coffee. The apparent reason being that all Fairtrade coffees from Latin America are from small growers. I explained that coffee had been marketed first as Fairtrade thirty years ago, and it had provided the means to improve the lot of highly exploited small growers in Latin America. Its success had then led to cocoa being introduced as Fairtrade as well. But when it came to tea, the Fairtrade Labelling Organisation (FLO) in Cologne faced a quandary. There were no small tea growers in India. Their dilemma ended, however, upon discovering Makaibari, with its strong record of corporate social responsibility activities. It was after studying the Makaibari model that the criteria for tea were set, and Makaibari was requested to join the FLO. Hence, large estates as opposed to small growers were FLO partners with regard to tea.

FLO has a set of separate parameters for producers and buyers. To become a member of FLO, both parties have to fulfill all the criteria and standards set by FLO. Once these are met, the producer becomes a member of the FLO register for producers and similarly a buyer can become a member of the FLO buyer's register. FLO has a panel of inspectors who audit the buyer as well as the producer annually in their respective registers for their respective compliance. Non-compliance invites a period for its redressal and a further inspection at a stipulated time-frame. Failure to comply within this time-frame leads to expulsion from the respective register. I also explained further that, to date, the FLO had not provided a single buyer, and it was Makaibari that had to convince clients to join the buyers' register.

Furthermore, the tea growing community was a completely different type of community to that of the coffee growers in South America based on the principle of statutory benefits for all under the Indian Plantations Labour Act. These included housing, negotiated wages, medical benefits, maternity benefits, subsidised food, clothes, children's education, fuel, annual bonuses, provident funds, retirement benefits, etc. The premium given by buyers went directly

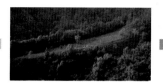

to the FLO in Cologne, which then remitted the funds to a non-governmental organisation (NGO) in India approved by both the FLO as well as the Reserve Bank of India. From there, the funds went directly into the account of the Makaibari Joint Body. This account is wholly managed by annually elected members of the seven villages for the social development programmes at Makaibari. An annual FLO audit and inspection is carried out to ensure proper monitoring of the funds.

Fairtrade farming

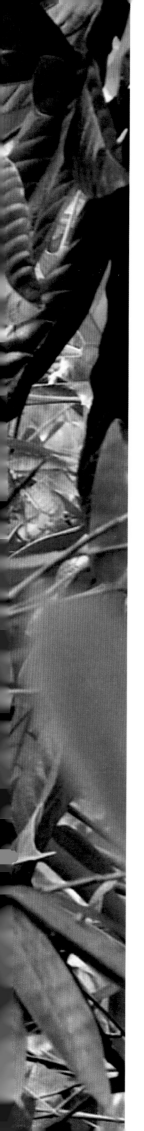

MAKAIBARI'S WILDLIFE

On the eastern side of Makaibari, there are thick woods – over 120 hectares of sub-tropical rain forests that are more than a thousand years old, and are home to the many animals and birds living there. The tea section next to it is called 'Coffee Khety' because my great grandfather had first planted coffee there in 1870 which my grandfather converted to tea; coffee then not being remunerative. It is funny to have some of the best teas from Makaibari named Coffee Khety ('Khety' in Nepali means field).

One day, my Plantation Assistant Dev and I were supervising fifty women, who were specially trained pluckers, picking superfine leaves for Muscatel Tea. Suddenly, we heard a strange whistling sound. All of us turned our heads and were hypnotised by a flock of Indian Pied Hornbills in flight over us. All of us watched in wonder as we beheld the truly majestic sight of parent hornbills teaching their newborn young to fly in formation.

at home in Makaibari

The Indian Pied Hornbill is known as the king of birds – a position similar to that occupied by the tiger among mammals. The female hornbill stays walled up in the trunk of a dead tree, her beak poking out of the aperture, with her clutch of eggs for up to six weeks. The male hornbill feeds her during this time, and the great bird loses her magnificent plumage completely. The wingspan of the bird is almost three metres, it has a half-metre long beak and a yellow crown on its head that is a half-metre in diameter. It does not twitter or sing but quite literally roars! It is one of the rarest birds in the world today and is almost extinct.

All of us felt honoured to have witnessed this sight. There are three separate flocks of hornbills in Makaibari comprising a total population of thirty. Sightings can be quite frequent but only if one is patient.

Within the woods of Makaibari, various species of insects, butterflies, birds and animals flourish. No one dares to hunt or even capture a butterfly as the entire community would immediately punish the offender. Hence, the forests are a haven for all life forms in danger of extinction.

There are three species of monkeys in the woods of Makaibari. The largest is the Golden Langur. With its black face, golden coat and two-metre long tail, it is an impressive ape. It is

rare species of butterflies and insects in Makaibari

the forest and the plantation in sync

a solitary beast, roaming the woods alone and pairing only during the mating season. The other species of monkeys in Makaibari, though not so impressive, are the Rhesus and the Assamese Macaque. Both species roam the woods in troops, with one large male as the boss and the others conforming to a strict hierarchical system of importance.

A war between two troops of Assamese Macaques is an awesome sight to behold, and a few years ago, I witnessed one. On one of my daily farm rounds (during which I normally cover 7–8 kilometres of difficult terrain on foot), I was surprised to hear a huge commotion coming from the Nayakaman woods in the western sector of Makaibari. Approaching the source of the noise with caution, I was amazed to witness the following scene.

The mothers of two groups of macaques were huddled with their offspring at a safe distance, while the young males screeched horrifically, chasing each other, tearing leaves from trees and seemingly raining blows on each other. I was unable to detect any real injury inflicted by a macaque from one group upon that from another. It looked more like a push-and-shove contest than any real attempt to cause pain to the other.

From my vantage point I continued to watch the scene while taking care not to disturb the macaques. Their seemingly war-like rituals raged for at least twenty minutes. After a while I realised what had caused the fight. It seemed that one group had accidentally encroached on the territory of the other, precipitating the war-like demonstration that I was now witnessing. Finally, the boss of the troop to whom the territory apparently belonged – an impressive male with a very red face and buttocks, walked up to the boss of the infringing group – an equally

impressive male, and hurled him to the ground with a resounding slap. Immediately, all noise stopped. The defeated ape picked himself up slowly and abjectly, made a few rumbling noises, gathered his troop and quietly left the site. Within minutes they were across the ridge a few kilometres away, swinging their way through branches towards their charted terrain.

Witnessing this fight between the macaques was a truly great experience for me. I was struck by the wisdom of their way of settling disputes with none doing grievous harm to another while asserting a territorial claim. It left me to wonder if human beings could also evolve to this state of coexistence with another who infringed upon his or her domain.

swinging his way through the woods

Elephants were sighted in Makaibari for the first time in February, 2003. Udayan, my elder son, was visiting Makaibari with his new bride, and we had set off for a long trek, going over the eastern part of Makaibari from Paila Khety in an arc all the way to the southern part via Varleni, the plateau to Chungey. Arriving at Chungey, our group was met excitedly by Dev

and two forest rangers, Jewan and Himul. They escorted us up a fairly steep slope on to the Chungey flat where Jewan pointed out the footprints of two elephants.

It seemed that two elephants had meandered out of the neighbouring Bawanipokri Sanctuary and up the steep Pankhabari Road, bringing traffic to a halt for over two hours. They had then slipped into the Chungey division of our estate. The pair had proceeded down the plateau, and upon arriving at the steep slope (that we had just climbed up) retraced their steps back to the reserve.

It is very unusual to find elephants in mountains as their size is cumbersome for climbing hills. And they had never been sighted in Makaibari before this, leading Jewan to comment, 'Our community God is Ganesh, the elephant-headed Deva of wisdom. He has visited us, in spite of the fact that elephants never climb mountains on their own. I am sure with their blessings we will really prosper.' Indeed, the elephants were a powerful omen as much needed rain commenced early the next morning and continued to purify and bless us all.

the map of Makaibari

MAKAIBARI FABLES

One day, I was at the western side of the farm, which is at a height of 1,000 metres above the sea level. At exactly five minutes past noon, I was in the Nayakaman section. It was the lunch hour between noon and one in the afternoon, yet I was surprised to notice that not a single plucker was resting after lunch before commencing work. I walked until one o'clock and was even more surprised to find all the pluckers returning very quietly to pick tea. But their eyes were smiling, so I knew that something was afoot. Their unusual bustle at the end of the break was strange. I had to find out the reason for this, but no one would tell me.

So I met the Area Supervisor and asked him to find out. Later, he quietly informed me that the workers had disturbed a panther that had killed a wild boar. They had then taken the wild boar, hidden it, and had trooped off to the woods to share it among themselves during the lunch break. Ruplal, the Supervisor, was indeed very angry that he had not got a share of the wild boar. Jokingly, I asked him to circulate an ancient Gurkha superstition that whosoever stole the prey of a panther could expect a visit from it, so beware! Of course, the tale spread like wildfire.

where inspiring stories are born

Next morning, I was surprised to see two ladies from Nayakaman. They had come to me for advice. The panther had actually visited Koilapani the previous night and had carried away a goat each from their pens. Both householders had heard the commotion and had stayed firmly bolted indoors. I suspected Ruplal's hand in the incident. However, I counselled the agitated women to offer prayers at the spot where they had snatched the wild boar, assuring them that this act alone would assuage the spirit of the angered panthers.

The next day being a Sunday, the entire village duly held a prayer ceremony which was followed by a picnic for all those who attended. It seems the panther's spirit was appeased for there was no further loss of livestock. The message was imparted that no one had the right to steal someone else's hard-earned food. Living in the lap of nature at Makaibari, it is elementary to respect the laws of nature – to work for one's daily needs. I smiled with content at the success of the solution – one that was simple but also had a deep reach.

End-October is a very special time of the year for everyone in the region. It is the time for the festival of Dussehra during which we celebrate the season's end by worshipping the Goddess Durga. It is an ancient celebration whose legendary origins are to be found in the Ramayana.

The Ramayana is one of the greatest ancient scriptures of India. The legend of the Ramayana tells us that after painful penances and prayers, Ravana, the King of Lanka, obtained the boon of eternal life from Lord Shiva. According to the boon, Ravana's life could not be taken by any man or Deva (God). Thus blessed, Ravana became a veritable tyrant, misusing his powers to such an extent that even the Devas became victims of his excesses.

To deal with the threat posed to them by Ravana, the Devas created Rama – an *avatar* who was neither man nor God. He was the perfect form of a human being, yet not a God. His wife Sita too was similarly incarnated. She was blessed with such beauty that it tempted Ravana to

the festival celebrations at Makaibari

steal her from Rama, rousing Rama to embark on a voyage to rescue her. Because Ravana was so powerful, initially Rama could not penetrate his defences. So Rama prayed to Goddess Durga for her help to overcome Ravana. During his prayers, Rama had to offer a hundred and eight lotuses to the Goddess. To test him, Durga hid one of the lotuses. When she pointed out to Rama that there were only a hundred and seven flowers, Rama immediately offered to shoot out one of his eyes with an arrow to make up the missing number and thus ensure the completion of the mantras. The Goddess was pleased and gave Rama the power to kill Ravana. With her blessing, Rama was able to overcome the evil Ravana; hence the celebration of Dussehra, to mark the victory of all positive forces over evil ones.

In 2001, the Dussehra holidays fell between 23 and 28 October. As it is traditionally the biggest festival in the region, the entire community was preparing to celebrate it with great pomp. As usual, there were to be two separate Dussehra pujas in Makaibari and Kodobari villages and I had been invited to both.

At two o' clock one morning, just before the holidays, security guards from the factory knocked at my bedroom window, awakening me from a deep slumber. They told me that

a burglar had been caught stealing teas inside the factory. I immediately informed the police who asked for a car to be sent to bring them to the factory. So I called my driver and sent him to the police station. Meanwhile, the guards together with other assistants succeeded in apprehending two more burglars who had been hiding inside the factory premises. The reason these two had failed to escape to safety through the windows was startling. The factory had been encircled by Makaibari villagers, armed with sticks and *khukris* (traditional Gorkha fighting knives) and with their hands joined such that not even an ant could escape. Over three hundred men and women effectively blocked all exits.

The police arrived in due course and were taken aback by the solidarity of the workers. As soon as the police entered, the entire mass of people commenced chanting in one voice, 'We want justice, we want justice.' The sentiment was clearly, 'Why can't the police come immediately? Why do we have to do their job?' This continuous chanting by over three hundred people in unison was enough to unnerve the police force, and their officer turned to me for advice. I suggested that he call for a larger force from the police station. Shakily, the officer did so. It was dawn when reinforcements finally arrived to escort the frightened burglars to jail and the police force back to the station.

It was a truly inspiring experience to witness the unity of the villagers in tackling what was a volatile situation so effectively when their temple — their factory — was threatened. I felt this could only have occurred with the awareness that jointly, they had the strength to overcome any adversity. The moral was clearly illustrated — united we stand, alone we are easily defeated.

Subash Sarker lives in the centre of Kolkata in West Bengal. His apartment is on the second floor of a house. His terrace is on the rooftop of his neighbour's house. One summer his children were having trouble sleeping because of the heat. It had been a long, hot summer, and the sun

Makaibari – a way of life

was strong on the rooftop. Mr Sarker thought of shading the roof with a vine – one with a lot of leaves. Hence, he planted a grapevine in the soil beside the house. The vine grew and grew well. It grew up the side of the house and onto the roof. Then Mr Sarker built a trellis on the roof for the vine to spread and provide shade, and thus, his family slept well even in the summer heat.

The grapevine provided the family with more bounty. It is over ten-years-old now. It produces grapes every year. Mr Sarker uses a third of the crop as fruit for breakfast; the rest, he squeezes into wine and vinegar. So he makes a lot of money from one grape vine.

The pruning litter and leaves from the vine are made into compost in a corner of the roof. He visits his local vegetable market regularly and collects all the peel to add to his roof compost in small containers, which he carries up one by one. He also has a vermicompost fill on the roof, as well as a pigeon pen. He mixes the leaves with pigeon manure. He fills pots and old tyres with soil, compost and manure mix. In it he plants tomatoes and other vegetables. Today, he has a beautiful vegetable garden on his rooftop.

At first, Mr Sarker had a lot of problems with theft; his neighbours would steal the ripe grapes. He solved this by growing cuttings from the vine and giving them to all his neighbours to plant in their houses and follow his example. Patiently he created awareness in the entire community. Now grapes are no longer sour in Mr Sarker's community.

Mr Sarker is the Chief Accountant at the Makaibari export office at Kolkata. He had been inspired by the resilience and innovativeness of Jamuni's success with the cow and biogas, the successful roadside diner established by Suna Rai's entrepreneurship and the vermicomposting techniques created and applied by the young men of Makaibari. The sustainable synergy released by these simple village men and women had awakened him to utilise the resources at his kitchen door to effect sustainable solutions in an urban environment.

This clearly shows that we can be sustainably creative no matter where we are.

Jewan and Robin are forest rangers who patrol along the main highway which divides Makaibari in two. The Pankhabari Road snakes through Makaibari along steep slopes with many hairpins and U-turns as it rapidly climbs from sea level to 1,500 metres in thirty minutes. It is quite hair-raising for first time visitors.

The road is extremely narrow in many places and uphill traffic has the right of way. However, during the summer when people in Kolkata desire a reprieve from the oppressive summer heat, they come up in droves to the Darjeeling hills. Taxi drivers throw caution – and rules – to the winds as they hare up and down the steep narrow road. This causes traffic jams. Hapless travellers munch snacks, and throw their litter of plastic out of their car windows as they wait. This creates a mountain of plastic litter along the otherwise pristine road. When their entreaties fell on deaf ears, Jewan and Robin decided to take more positive action.

At one of the weekly ranger meetings, a visiting World Wildlife official had narrated the story of an antelope which had swallowed a plastic bag that had contained salted nuts and had died. A visitor had carelessly thrown the bag, which had clogged the antelope's intestine, causing death. He warned all the rangers to be alert to the potential for such incidents particularly along the road where the deer population was high.

Robin surveys his domain – Pankhabari

Moved by the story, Robin and Jewan collected all the plastic litter on Pankhabari Road. They then made tubes out of it and have grown indigenous plants in them for reforestation at Makaibari. Once the saplings are planted, the plastic is sent to the recycling plant at Siliguri. The Makaibari Joint Body has awarded a special prize to both the rangers for this innovative positive work which has benefited the environment enormously.

When one cares, one finds solutions for the benefit of all. Networking these initiatives would result in a tidal wave of sustainable renaissance globally and make the world a safer, healthier

place for the next generation. I hope this story about Jewan and Robin is inspiring to you. After all, self-help is the best help.

One day, in Makaibari, it had rained in the evening around five o'clock. The day had been hot, and the evening showers brought the temperatures down considerably. There was a pleasant breeze, as the moon emerged and the hills reflected the silver rays of the droplets caught by the trees. It was indeed awe-inspiring to be a part of this magnificence. I looked across the hills and thought that I was indeed blessed.

Uplifted by this experience, I entered my house, unmindful of the fact that the front door was ajar; normally it is shut in the evenings to keep all the insects out. I immediately noticed a large frog, sitting in the middle of the corridor, looking at me balefully with its large eyes. I called out to my servant to fetch a basket for the frog to be trapped and released in the nearby woods. He came running with the basket, and then froze. Tremulously he stammered, 'Sir, there is a large snake behind you.' I turned my head and, lo and behold, there was a metre long brown snake, coiled, with its tongue flicking and it was looking at me angrily, as I had disturbed it from its prey, the frog. I indicated to Dilip to pass me the basket which he hesitatingly did. Slowly, without any untoward movement, I dropped the lid on the snake, placed a board underneath it, and carried it off to the forest where I released it much to Dilip's relief.

The frog, meanwhile, had hidden behind the table leg and with much prodding I succeeded in transferring him to the same basket whilst Dilip still stood motionless in shock. Finally, a prod in the belly stirred the frog. Without further ado, he snatched the basket from my hands and went outside to release it.

On his return, Dilip asked me, 'Why aren't you frightened of snakes?'
I replied, 'I am always captivated by nature's ways at my doorstep; so much so, that I am

unaware of any negative fallout.'

'Sir, most people, including me, would have been inclined to take a stick and kill the snake, but you caught it and released it. Even when it was in your house! Why?'

'I respect all life forms, and if we learn to live in harmony with the vibrations of nature, all our lives are sweetened. Tell me Dilip. Are you frightened now?'

Dilip said, 'No sir, I am thrilled to have been a part of this beautiful experience. I shall in future always cherish the beauty of life.'

In the next few days, Dilip nursed a wounded, wild, green pigeon and when it had healed and recovered, the two of us ceremoniously released it to be free. Dynamic interaction is to liberate one another from the shackles of our minds. We smiled at each other knowingly as the beautiful bird soared away to the deep woods.

the cow – instrumental in sustainable practices

THROUGH THE VISITORS' EYES

❀

Blake Rankin is a Makaibari supporter from the west coast of America. He has a very successful tea and herbal company called Granum. Blake was a Zen Buddhist in Japan for many years. He gave up being a monk to marry a very beautiful Japanese lady and returned to USA. Unfortunately, after they had two children, the couple's marriage ended in divorce and Blake remarried. Both Blake's families live close to each other, and he enjoys the company of all his children – as well as selling Makaibari teas.

When work and life pressures became excessive, Blake visited Makaibari to relax and recharge himself. He had great spiritual empathy with both the land and the people here, and healed rapidly.

On his last visit to Makaibari, he smelled the soil and declared, 'This is land fit to sleep on!'. Immediately, we asked him to prove his word. Blake spent the night in the tea plantation with the lurking threat of poisonous king cobras, panthers and nocturnal visits from other animals. He was not bothered and slept the night on the ground, awakening very refreshed the next morning. We were all amazed. Mulching was the simple act that had created this miracle. Blake's

homestay at Makaibari

experience was the ultimate affirmation of the statement that healthy soil means healthy mankind. Everyone at Makaibari respects and loves this great man who showed us what a special place our efforts have created – a place where all can coexist harmoniously.

Ulrike Mueller visited Makaibari from Germany. She was twenty-two-years-old at the time and a student at Hohenheim University in Stuttgart, Germany. An exceptional girl, she lived with the Makaibari community for three months to create a herbarium with local herbs. She has left this for future generations at Makabari to use; specifically, to enable us to use traditional herb perparations as an alternative cure to allopathic treatments. She is unique to have possessed such wisdom for one so young. We remain captivated by her ability to sweeten our existence. Here is some of her story told in her own words, a few months after her arrival.

I can clearly remember the feelings I had on my flight to India, which is now more than one month ago. There were many uncertainties like the different culture I would have to cope with or the possibility of feeling homesick and lonely. Besides, the question whether my idea of collecting medicinal plants in the forests of Makaibari would succeed, occupied me. As I think that it is very important that the knowledge of the old village people shouldn't get lost, I had decided to look for local herbs and ask the villagers about their medicinal qualities. I wish younger people would be more interested in traditional medicine as it doesn't have any side effects. I planned to press the plants, make a herbaria and try to find out their local and scientific names. I hoped that the result of my work would be useful to many people.

Soon after I had arrived at Makaibari, I realised that there was no reason to be worried. People welcomed me warmly and I knew that there were several people whom I could contact if I had any problems.

When I walked through the forests of Makaibari, I was again and again impressed by the enormous biodiversity of plants and animals there. It was really a fantastic feeling to be at some of the remote

places like in Nayakaman or Chungey where the only sound you can hear is that of rivers, crickets, monkeys, and birds. Besides, I also got an impression of the life in the villages.

People were always friendly and open-hearted – they liked telling me about their religion and traditions. Then, there were the discussions with Mr Banerjee about the meaning of life or the philosophy of organic farming which often gave me a new way of looking at things. Moreover, it was very interesting to meet people who come from all over the world to Makaibari.

I think Makaibari is a unique place in many aspects, and I am glad I could experience its atmosphere from so many different points of view.

Ulrike Mueller

Melisande was an unusual German girl who visited Makaibari in her gap year. This is her personal experience, in her own words.

My stay in Makaibari Tea Estate: 28 September to 3 November, 2001

I'm Melisande Rodenacker, twenty-years-old and I live in Germany. I visited Makaibari Tea Estate for about five weeks and have tried to write down my experiences in order to share them with you.

My first days in Makaibari

'I love India' used to appear on almost every piece of paper among the drawings I did during the rather boring lessons in grade nine. In 1996, I had come to India with my family. We spent two weeks in Kolkata, Shanti Niketan and in Digha. This trip had given me great moments of happiness and love, and a good friend in Kolkata, Chandrima. When I came back to India and arrived in Makaibari, I first got offered a cup of tea. It took about nine days to get to know the new place and to get used to it. I got

homestay at Makaibari

introduced to the people and to the way of living, working and cultivating in Makaibari. I got some general idea about the plantation, since I visited its different divisions.

I met blood-sucking leeches, toilet paper eating snails and mice that slept in my bed. During this time it was raining very heavily; it was cool and wet. Clouds and fog hid the surrounding nature. It looked very magical; an endless play of veiling and unveiling nature's beauty.

Only gradually was I allowed to see the more distant beauties. One day I was able to look down in the valley, the next day I could see the houses of Kurseong on the ridges of the mountains, then there appeared the mountains across the valley and I was very stunned when I could see the plain outstretched like an ocean far below with the rivers meandering through it.

I wanted to experience the life in Makaibari, so I asked whether I could do some work. I had imagined something like what we in Europe call 'praktikum' or internship, but I soon realised that it wouldn't be possible to do this in India. I couldn't be much of a help because there were always lots of other workers who could do the same thing much better and faster. And it was unusual enough that an European girl did some tea plucking, planting and weeding.

For a few hours on three days, I went tea plucking. One day, I went to plant tea and the next day I did weeding and mulching. I was working in different divisions with different groups of workers. I was always welcomed and the ladies helped me a lot. Especially the first day, I must have looked somewhat clumsy. The whole group watched me and laughed a lot at my attempts in a very nice way. I slipped a few times, missed the basket when I tossed the tea leaves back and once, even my basket fell

down. They showed me how to pluck tea, how to put on the apron, the umbrella and the basket, which is carried with a headband on the back, an unknown technique in Germany. They rolled up my sleeves, put back my hair which kept falling into my face, gave me a plastic foil to put around my waist to keep my pants dry, they showed me which bushes I should pluck and where I should go. When it was a little steep, they held my hand. They hardly spoke any English and I didn't speak any Nepali, so it was communication through smiles and gestures.

I enjoyed it very much. Most of the time I had to concentrate on what I was doing and on keeping my things together. It's not so easy to move with a basket and umbrella contraption between tea bushes in the mountains when you are not used to it. Sometimes I just enjoyed the feeling of working in this group, in the rain, breaking off the young and beautiful shoots, feeling the strong bush against the legs and the soft ground under the feet. One day, part of a group gathered around me, broke a betel nut into pieces, every lady got a piece and said her name; then we were friends.

In the divisions where I plucked tea, people first observed me but after a little while they got back to their work. When I went to do the planting in a different division, one lady helped me while the rest of the crew

tea plucking

watched me all morning. It is a strange sensation, when you are being observed like this. But I also enjoyed planting. It is something very special to touch earth and plant a bush which will live for about eighty years, producing a lot of new tea.

In Makaibari, the earth is soft and smells good. There are a lot of earthworms. When I finished and the earth on my hands dried, they were of a nice red colour and shimmered in the sunlight due to the mica. It was a beautiful moment when I held them under the fresh and clean water of a rivulet running through the plantation. For me, this work was a pleasant experience. For the workers it's hard work, especially because they have to do it day in and day out for years. To do my work, I depended on the help of the Assistant Managers who were very nice and took me to the groups of workers, introduced me to them, explained to me in English what I was supposed to do and took the whole responsibility of my visit. I am very grateful to all three of them; they really took good care of me.

the basket and headband

In the mornings, before I could go to the plantation, and also sometimes in the evenings, I had to wait for the Assistant Managers or Rajah Banerjee in the office. I spent a lot of time sitting in one of the rooms, watching the men do their work or chatting with one of them, drinking one, two or three cups of tea (I mostly drank green tea and in the plantation it's not easy to find a toilet, so my bladder got trained to hold a lot.). I had enough time to experience the office. The most surprising thing is, that all the 'paperwork' really is paperwork, everything is properly written down in one of the many books piled up everywhere in an incomprehensible order. The only job of the computer seems to be the use of the electronic mailing system.

I found Makaibari to be a very pleasant place. I was treated very well and was given a lot of attention. Although I was a little sick during the first week, I was happy, I enjoyed my stay and am grateful that I was given the chance to meet such wonderful people.

Nowadays, people believe that it is impossible to change something towards the good. Most people of my age think that things have to be changed but they have lost the motivation and hope to do so. Makaibari is an example that good things can be successfully created. There, a lot of dreams, ideals, wishes of a right, balanced, just and fair life with society and nature have become a reality. And quite a lot of people have found the purpose for a peaceful and dynamic life without violence, hate or extremism. People can love and respect each other and themselves much more there than in other places on Earth.

The days before the festival

One evening, in the second week of my visit, Mr Banerjee, the Assistant Managers and I met. I talked a little about my experiences in the plantation and then they thought about what I could do in the following weeks. One of them had seen that I know how to draw. So my new task became to draw, paint and sketch. I was pleased that from now on, I would work independently and on my own responsibility. I was asked to decorate one of the maps of Makaibari which shows the outlines of the areas cultivated with tea and the villages in the plantation. So for the next one and a half weeks, I was busy copying lots of birds and animals from books on the map, filling the blanks with plants and simply covering the whole big piece of paper with colour.

It wasn't a tremendously interesting work and it took me very long. I really wanted to create something beautiful. While I was working, lots of people came in to admire the map. I was thankful for their cheering me up. The final product looks nice, although some things could be better. The day I finally finished the work was the beginning of the festival season. The same day, I saw my first Tea Deva, a truly magical sight, and Mr Banerjee came back from Kolkata with the needed money to pay the incentive bonuses, which meant the end of some difficult days.

Tea business had been bad that year and most tea estates didn't know how to pay the bonus. The workers went on strike. For about a week, people were worried and the atmosphere was very tense. So when the money finally arrived, everyone was really relieved and the air was filled with laughter and jokes. I had never been so happy seeing someone counting money. From that moment on, everyone got into the holiday mood, everywhere the preparations for the Durga Puja were going on. The rainy season was really over and the next weeks were going to be warm and sunny. The views amazed me everyday and flowers burst into blossom everywhere.

I was planning to do some sketching outdoors, portraits and village scenes. I didn't know how to proceed since I had never done this before and I couldn't imagine how to explain to people what I wanted to do without speaking their language. But when I started, it turned out to be quite easy. Nobody misunderstands a blank piece of paper and a pencil, especially when I showed some finished drawings. So I started with the portraits of the two old men who worked at the guesthouse as guards. They turned out good and I wasn't to worry about what to do during the festival days.

When those days approached, me and Uli, a German student who was also staying in Makaibari, didn't know what was going to happen and what we would do. Each one of the men working in the office we had had the most contact with, was going to spend the holidays with his family in Siliguri or Kolkata. But then, everything turned out a little different than expected. During the following week, I was busier than ever before in Makaibari. I didn't even get to do any drawings.

One day, the Banerjees took us and some other guests to Darjeeling. We went to the Planters' Association, a club for high society and tourists. The place and people there seemed to be quite influenced by the British heritage. Then we were treated to a delicious lunch. I wasn't allowed to finish before my belly wouldn't take a last grain of fennel anymore. After some minutes being stuck with the cars in a street only allowed for pedestrians, supposedly being stopped by an angry drunk man, surrounded by a crowd of curious people, we were saved and could proceed on our way to meet the princess of the Maharaja dynasty in her summer palace. There, we were shown the house, sat a while and had a chat and then we drove back the curvy and bumpy road to Makaibari.

Durga Puja

The Puja began the next day. In the Makaibari Tea Estate, there were two pujas in two different villages, one in Foolbari and the other in Makaibari. The first morning, Uli and I went to Makaibari to have a first look and to find out what was going on.

A temple-like construction of bamboo and cloth had been made and it was decorated beautifully. Inside this there were statues of the gods, also covered with decorations. In front of them, oil lamps and offerings, mainly of fruits and flowers were kept. In a circle around it sat a priest and some young people praying. Sometimes they turned on the microphone and then the prayers could be heard all over the valley. When we arrived, we were immediately offered chairs on the nearby stage, we talked to some girls and boys of our age and to some children, some of whom spoke perfect English. We also got a red patch of rice on our forehead, a banana leaf plate with sweets and fruits and a cup of tea.

the Durga idol in Makaibari

At noon, we went back to the guesthouse for lunch. In the afternoon we went to Kurseong to write emails and then we visited another Puja. We just happened to walk in and we were served another meal. So we had to eat again and it was delicious. It is an unusual situation when all of a sudden you sit in a hall, in front of beautiful statues of Gods, eating delicious foods with your fingers on a banana leaf.

After a nice walk up the hill to places we had not seen before, we went back to the guesthouse. The sun had just set and we couldn't turn our eyes away from Kanchenjunga. The snow-white peaks reflected the colours of the sunset. It was an amazing spectacle.

Then we had dinner. Usually, we were told that Durga Puja was a time of fasting. For me it was the opposite. After dinner we went to Makaibari because there they had a cultural programme. The whole

village gathered in front of the stage. Children sat on the ground, behind them on benches were the ladies and more children, and the men were at the very back. Everyone was laughing, chatting and snuggling up to each other. Uli and I, as guests of honour, got the chairs. On stage, the preparations had not been finished, young men were moving around fixing the microphone, the music and the decoration.

When we thought everything was finally ready for a beginning, they started whispering and organising again. We didn't realise that it was due to us but then two men came down and asked us to come on stage. That was quite a surprise but then we stood there on the stage in front of all those people. Next to us were three little girls, beautifully dressed up in traditional Nepali dresses. They were to give us the 'khada', the small, light-yellow silk scarf, to pay us an honour. When the littlest girl put it around my neck, I had to bend all the way down to the floor, because I'm so tall. The whole village had a good laugh. I am even taller than most of the men there. I was told that when they talked about me in Nepali they called me 'the long one'. Then we had to light an oil lamp, it was some kind of an opening ceremony. I even found some words to say to everyone which weren't too stupid. But I don't think too many understood because of the language and the microphone quality.

The following programme was really amazing. There were many solo dances of different styles, modern/western as well as Indian/traditional. Girls and boys of every age danced and I was stunned because they were so good. I had never seen anything like that; so many young people dancing with joy and passion. Then there was also some singing; I wish they had had a better microphone.

The next day Uli went to Foolbari and I stayed in Makaibari. In the morning, there was a big ceremony going on and Mrs Srirupa Banerjee and I, as guests of honour, had to participate actively. The ceremony was very beautiful. In Hinduism, ceremonies are very complex and there are always some instructions on how to do everything. So it wasn't a problem that I didn't know what was going on. Although I didn't understand anything, neither the language nor the symbolic meaning of all those actions, I enjoyed the experience. It was very colourful. There was singing, drums, prayers and various actions.

In the afternoon, I went to Foolbari for a short visit and at night I was back for the next cultural programme. This time there were group dances, theatre sketches and a game. I have no words to describe the dance and the acting. The Makaibari youth is brilliant. I sat there, between my new friends, watching the dances, enchanted and laughing at the acting although I didn't understand a word.

The game was funny too. There was a closed box with chickpeas in it and everyone could pay ten rupees and take a guess on how many chickpeas there were. Then, in the evening they counted the peas. If someone guessed the right number of peas, he or she would get all the collected money, if nobody guessed correctly, the money would go to the organisers of the Puja. The one who would get the closest to the right number would win a chicken. A girl, who didn't dare to take it, won the chicken. It was funny watching the men counting nine hundred and something peas on a stage and the chicken looking down on the audience suspiciously.

In the end, the young people danced, they also asked me to come on stage and dance with them, which I did. Unfortunately I couldn't present any solo dance, I would have liked to, but I had never danced as much as they did, in that style and with that kind of music. So I just copied the people around me. For me it was just enough to make me feel like I should learn how to dance, but everyone seemed to be happy.

The next morning was the final day of Durga Puja. I joined in the procession to take the statues of the Gods to a river and immerse them into it. My friend Barshat convinced me to let them dress me up in the traditional Nepali dress. Everyone who saw me was first surprised and then happy to see me like that. Most of my friends also wore traditional dresses or a sari.

The statues were on the trailer of the tractor which was decorated. They blasted Nepali music and we walked in front of the tractor. The guys danced. Girls were more reasonable, because even without dancing, it was tough enough to walk uphill to Kurseong in the warm sunlight. I enjoyed it very much, walking with my friends, with the music, as part of the group through the beautiful mountain roads and then

through Kurseong. Every once in a while we stopped and a group presented a dance. But also while walking, there was dancing and clapping hands. When we reached the river, the statues were submerged into it and then everyone started to splash water at each other.

The guys had to walk back, probably after hanging out a little in Kurseong. We girls had the privilege of being taken on the trailer of the tractor. It was a fun ride. We sat very tight, almost on top of each other, and the group was singing a lot of songs at the top of their voices. The toy train was passing, I looked up, right into the faces of two western travellers. I had to laugh; they had very long hair and beards. You meet a lot of odd western tourists in this area. But then I thought how they were probably surprised to see a white girl in Nepali clothes sitting with singing girls on the trailer of the tractor. A little later two guys jumped up on the trailer, one of them was really drunk and he then fell in love with me. He started to talk to me in Nepali and English, telling me about an arrow in his heart. He

Makaibari in full bloom

repeated the same sentence over and over again until the girls cut him short by singing whenever he raised his voice. It was a funny situation.

My last week in Makaibari

Now the festival was over. People told me to come back for the next Durga Puja or at least stay until Diwali, the next festival. But my last week had come. I made a few more drawings, but I spent most of the time with friends.

Barshat, the girl who had been with me most of the time during Durga Puja, became a good friend. We often sat together and talked. I think and hope that we'll stay in contact for the rest of our life. One day, I asked her whether she wanted to come with me for a walk down into the plantation. She came and we spent a wonderful morning walking. We saw a lot of beauty and visited some tea plucking ladies who remembered how I had plucked with them a month ago. When we finally returned we were really tired, but happy.

One day, two girls from a different village, who Uli had got to know, took us to Kurseong. We went to Eagle's Craig, a memorial for the Gurka fighters and a look out point. From there, you have a great view down to the plains as well as to the Himalayan Mountains. We strolled around in Kurseong, holding hands, got snaps clicked by a photographer and then we went to watch a Nepali movie. It was nice; I liked it although it was very long and quite simple compared to what we are used to in the West.

The day after that, I wanted to go to meet Subu in Koilapani, a village a bit farther away. I had been introduced to him in the very beginning of my stay and I was told that he is a musician but I had not heard him play. This visit turned out to be very nice. Subu and his father Ruplal played some songs, then I played some European songs on the violin and the guitar. We also played a Nepali song together. Music doesn't need a common language, it is international. Subu's sister played some cassettes with Nepali music and danced, I danced with her and her friend. Then I was invited for lunch. It was delicious. Later on, Subu and his brother took me for a round to show me the village which also was really nice. I got to see a lot of things.

To the European eye it looked very simple but people seemed good and strong. They hold together and have a living tradition. Although they have such a tough life with a lot of hard work they seem happier than people I observe in Germany.

On my last day in Makaibari, I went down to the village to meet Maita Sing Rai, an old man I had sketched a portrait of and now I wanted to give him the drawing. Someone I asked showed me the house where I met him and his family. I was invited for a cold drink and cookies. We chatted a bit, some minutes later Biswas walked in and we were quite surprised to see each other. We had met at the festival. He was with me pretty much the rest of the day. He and Brawesh came with me to my friend Barshat's house.

In the evening, they took me to the place where they have cultural meetings, dancing classes etc. and, surprise, surprise, the whole youth of the village and even some older people were gathered there. They gave me a farewell; I think the nicest farewell I ever got. I was greeted and given some really nice gifts and then they danced. Once more they danced with the songs we had heard so often during the festival and also to some other music. I was so happy to see all those familiar faces again and those wonderful dances full of rhythm, movement, passion and joy. During some dances I tried to dance with them. When it was finally time to go, I tried to look at everyone once more and it was so moving, what I saw in their faces.

I had not even talked to most of them and yet, they had become friends. The most important communication with those people had not happened with words. I had lived with them, had participated in their festival, I had shared their joy and I grew to love them and they obviously did too. One of the girls had noted down in a book that friendship is like a rainbow between two hearts. I really like that picture. Just that there weren't just two but many hearts.

During the last weeks I spent a lot of time with my own generation. I enjoyed spending time with them. They did not smoke or drink alcohol but they danced, acted, sang and organised this great

festival. They seemed to treat each other in a very kind way. I respect them for how they manage to get a balance between the traditional Nepali and the modern western lifestyles. Things are changing in India and I'm very happy that there is a youth that is able to live both to a certain point, a traditional as well as a modern western style.

The last morning, I said good-bye to the men who work in the office and the factory, to Rajah Banerjee, to the guards, the ladies who do the cooking in the guesthouse and to my friends Barshat, Dev and Uli. I had my last look at the landscape which had become so familiar and then off I went, down to the plain, to the station and back to Kolkata. It was sad and it took me two days to get used to hot, dirty, noisy Kolkata. But soon I had found new friends, and spent time with Chandrima.

I would like to add some words about some of the great people I spent time with in Makaibari. There is Rajah, the owner, manager and creator of what Makaibari has become in the last decades. He is a bit like a locomotive engine of the train. He seems to be loaded with energy, ready to give lectures about everything, encouraging and inspiring everyone he meets. I spent hours in his office, drinking tea, listening to what he had to share with me, watching how he received his employees and did his work. Even when I didn't understand what he told others, (because he spoke in Nepali or Bengali) it was entertaining to watch him and see the reactions of the listener.

In the beginning, I went to the plantation twice with him. He showed me the tea, birds, plants, butterflies and how every spot of green is different. Rajah is a great person, hilarious and serious, loud and sometimes silent, very open and honest, most of the time he speaks out what he thinks frankly, needs company and needs solitude. He seems to have a deep understanding and a good perception of what is happening around him.

Another human being I am very thankful that I could meet is Dev. He is an Assistant Manager; he became a very good friend of me and Uli. Thanks to him we got to learn a lot about India.

After getting to know him, I realised how much strength and independence is needed to do what you feel is right and not follow others blindly. It is difficult; one has to be much more alert and has to work hard. He impressed me with all the good intentions and love he has for his surroundings.

Then there are the people who looked after me and the other guests in the guesthouse. The four ladies – Radhika, Sabita, Mona and Lalum, and the two old guards. The ladies cooked three delicious meals every day. Sometimes I joined the ladies in the kitchen. I learned some things about Indian and Nepali cooking and it was always really nice to talk to them. They spoke a little English. Often, I underlined what I said with acting and that made all of us laugh. We usually laughed a lot together.

The guards were also very concerned about us. The older one of them, Tshong, showed me places around the house. He enjoyed teaching me a few words of Nepali. Once, he brought little chocolates for me and Uli. When I had gone to the town and was a little late in returning, he was already coming to meet me, worried where I was. He took me to his house to show me his room full of Buddhist gods, religious items and decorations. It was all very beautiful.

During the festival, he woke me up early one morning, took me to the place where you can see the Kanchenjunga, invited me for a delicious breakfast and took me to a friend's house in the village. He hardly spoke any English but somehow I usually understood what he wanted to say.

I met some other Nepali village people. Unfortunately I couldn't talk to the older ones. Whenever someone translated some things they said, I found some kind of natural wisdom. Like one old man; I sketched a portrait for him, very quickly, because the light was fading fast. I wasn't happy about what I had produced. I don't think he is recognisable in it, but he was very moved and thankful. Later on someone translated what he had said, that this drawing is real, unlike photographs which are artificial.

To conclude this tale about my stay in Makaibari, so abundant in words: I enjoyed every day and I was very happy experiencing this new place and meeting those wonderful people. I learned the kind of

important things that are never taught in school. Everyday, some small or big thing surprised me and surpassed my expectations and imagination. There were a lot of adventurous moments and beautiful sights. I am very thankful that I was given the chance to experience all that.

I wish you and everyone all the best.

Melisande

Felix Zeller was a twenty-one-year-old student of psychology at the University of Leipzig, Germany, and a very intense young man, when he came to stay at Makaibari. His father is a renowned politician in Bavaria while his mother is a judge. He spent a week in Makaibari in 2002. He is an astonishing lad and I was amazed at the wisdom he had for such a young person. We all learnt a great deal from him. I feel that young people today are far more spiritually inclined than they were when I was twenty-one. To all the younger generation, I salute you for your perspicacity and wisdom. The future indeed is very bright.

Here are Felix's thoughts, which he gained after living with the community.

The tree is a symbol. The tree is life developed from a seed. A seed is closed within itself, is a circle, a universe of its own, containing all the wisdom of nature, waiting to be released; waiting patiently for the water and soil to awaken it. The sun falls on the watered soil and the seed begins to grow. It desires to reach out to its life giving source – the Sun.

The trunk is the base of its concentrated power, its sustainability for terrestrial existence. Over time, to attain a sense of its environment, it sends out branches, to feel what is without from within. The branches

are the sensors. The leaves, blossoms, flowers, fruits, and seeds are an extension of this reality. The foliage will hide the trunk, but the truth will remain essentially unaltered.

However, with the foliage and lack of Sun the trunk will grow thinner with the distribution of the branches. The focus shifts to the crown and top. The tree thus finds an end in itself, without reaching the sky, a circle of content completion; a complete life cycle of its own.

What then is the purpose of breathing and reaching out, as it cannot reach the Sun? Transforming soil, reflected in its appearance, making it visible to others and ultimately building a basis for a new life via oxygen. Moreover, the decay of its leaves and twigs on the ground is also concentrated energy, is knowledge for the foundation for a new life.

the heart and soul of Makaibari

The seed that falls from the tree on to this prepared ground is part of universal knowledge. New trees emerge out of this ground, transformed, and develop into new light. Knowledge transformed into energy, energy is motion. Thus the value always lies below the surface from which we evolve.

The spirit, the essence of life, will not meet the God in eternal heights unless it repeatedly falls down to the ground again and again, till its potential energy is converted to synergy. Hence it is essential to realise living in society is symbiosis of multiple impulses which each of us must channelise like the tree.

Thus, knowing others enables an awareness of oneself. Thus the aim of society is to make every individual aware of this salient ethos of unity in diversity.

At Makaibari, there is a profound sense of peace; this enables the space for self-respect to be attained as per every individual's time and space. Respecting oneself, means respect for all beings, so a beautiful tapestry of harmony can be built living with natural laws. The measure is the creative output. One sees, smells, and tastes the tea – the focus of the community, as its tool of sustainable evolution. It is wonderful; it is art with a purpose, the end of a long chain of values. The harmony of the tea is the reflection of the harmony of the community!

I feel honoured to have made this self-realisation from the trees of Makaibari.

Felix Zeller

Below is the story of Jean Phillipe Noel, a young Frenchman who was seeking alternatives for his life and arrived at Makaibari after a year spent travelling around the world.

After eleven months of travelling around the world and after having visited so many Fairtrade organisations globally, a combination of circumstances led me to the Makaibari Tea Estate. A few years ago, I'd have said it was because of simple causality that I've had come here. But right now I don't think I can leave it to causality. I arrived at Makaibari because it was meant to be so. On the other hand, taking a much more personal point of view, I came here because I had to learn things from Makaibari and its people. From all her people – from pluckers, guards, or servants to the big boss, Mr Banerjee, who I would rather call the father of the whole community instead of the big boss, a term more suited for a conventional company.

In conventional companies, the only target pursued is 'flavour in the balance sheet'. Makaibari is not only about economics but also about environmental and political consensus. Here, nothing stands alone, everything is interconnected, everyone in the community, every single insect and every single tree is essential and is contributing to the whole sustainable development. The term sustainable development comprises many different aspects. This is what is amazing about Makaibari. This organism perfectly illustrates each and every aspect of sustainable development, holistically.

I had not scheduled to be at Makaibari for so long as my original itinerary was overloaded. I had intended to rest for a few days at Darjeeling, and move on. I stayed on and on, riveted by the magic

profound peace

of Makaibari Nevertheless, one needs to spend a considerable period at Makaibari to imbibe the deep essence of this place. That is why, after a few days, boarding the train at New Jalpaiguri station for Varanasi, I felt as if something was wrong, was incomplete, and I returned to Makaibari. Each day was a revelation of learning and had I continued further I would have learnt more. I am continuing to learn although I know now that the entire magic of Makaibari has not been revealed to me. This is actually very good, for this will give me a very good reason to return, again and again, till I find it.

Jean Phillipe Noel

Professor Gunther Faltier has founded a mail order company in Germany which simply has no parallels. He only imports Darjeeling vintage teas – sustainably grown. It is under the versatile stewardship of Thomas Rauchle – the CEO of Projektwerkstatt Teekampagne, Berlin, that this unique supporter of Darjeeling has become the largest buyer of Darjeeling tea globally within a decade – an unparalleled testimony of Thomas's honesty, integrity and commitment to Darjeeling. Thomas is Darjeeling districts' best friend. Here is the Makaibari experience of Thomas Rauchle.

The end of March 1993 was my very first time in Darjeeling. After landing in Bagdogra, we drove directly to Darjeeling and I was fascinated by its unique landscape and all the spectacular impressions.

I have visited a lot of tea gardens and took the chance to pick up all the background about plucking and manufacturing Darjeeling Tea. I was amazed to witness the steepness of the hills and enjoyed the thrilling roads of Darjeeling.

In the very early morning we could see the rising sun which submerged the Himalayan range and the mind-blowing Kanchenjunga in a golden light. Down from the valley, we heard the prayers and singing

of the monks. Aside from all the fantastic impressions, I remember very well that I had a feeling of missing something; it was a little unreal but I was unable to point out what it was.

After a couple of days, we went southwards to our last station of the Darjeeling trip – Makaibari – even the name of this tea garden sounds a little mythic. I was a little sad that it was impossible to see the snow-covered mountains of the Himalayan range from Makaibari but I was more than surprised.

Invited by Rajah Banerjee, the owner and a pioneer of organic farming, we took the chance to get to know the nature of Makaibari which is probably the oldest bio-dynamic tea garden. How can I describe the feeling? It is as if you look behind the curtain and you discover a new planet!

Immediately, it was clear what I had missed in the days before. Wherever I looked, I saw little insects, flies, bees, worms, variations of butterflies, thousands of spider webs and many more. I think it was the first time that I heard the singing of birds in Darjeeling. Of course I have seen some insects here and there in other tea gardens but never this diversity of species. I will never forget the squadrons of dragonflies I have seen – hundreds!!

Wherever I was – I had the feeling of receiving some special energy – a feeling that I had never experienced at other places. In one of the jungles of Makaibari, I heard the sound of the river, saw rare plants and the traces of a lot of wildlife. Unfortunately, I could not spend more time there but it is reason enough for me to come back to Makaibari as often as possible.

In the following years, I visited Darjeeling several times and I always took the opportunity to spend a little time at Makaibari – I wish it could be more – and every time Rajah showed me something new. I would love to trek through Makaibari's deep forests, the tea fields, the jungles, to pass the villages and to enjoy the spectacular view into the plains.

In 2005 when I was visiting Makaibari with some colleagues, we were told that some Himalayan

panthers were in the area. We decided to go out in the night to the places where these panthers were normally seen – accompanied by an experienced guide of Makaibari.

After a thrilling hike, we came to a place which was fantastic. It was a full moon night, a clear sky and we were sitting on the ridge, enjoying a wonderful view into the plains down to the lights of Siliguri, some owls were crying through the night and the whole world was illuminated by the full moon. What a feeling!!

After a while, our guide gave us a short signal with the torch and we saw a Himalayan panther drinking from the stream, not more than perhaps 30 metres away from us. His eyes were blinking out of the dark like diamonds and it was my first experience of a wildlife adventure like this. After a couple of minutes, the panther went back into the jungle and we turned back to the fantastic view into the plains.

A year after the incident, we saw the traces of a tiger whose habitat was on another side of Makaibari. Needless to say, we tried to have an adventure again, the way we had had with the panther but this time there was no luck.

Since I have visited Makaibari many times, walked up and down the mountains, I respect the work of people in the tea gardens. All the varieties of Darjeeling Teas are handmade. Yes, the producers use machineries for manufacturing the tea but quality starts in the field and plucking in Darjeeling can not be done by machines. The character of Makaibari tea is something special, nobody can copy it. The unique combination of nature, the weather conditions and the ecosystem reflects in the tea.

And not only this, the families who are living in the villages, which are a part of the Makaibari society, are all fully integrated in the system; a lot of social projects have taken place there. Makaibari has a computer training centre for kids to make the younger generation fit for the future.

Needless to say that Rajah Banerjee and his lovable wife became very good friends and I wish to have the opportunity to go back to Makaibari very often.

Spending some time at Makaibari gives you a feel of the 'Makaibari Spirit'.

Once you have it, you will never forget it!

But be careful — it is addictive.

Thomas Rauchle

Kiran Tawadey has been a Makaibari associate for long. What follows are her words and feelings about Makaibari.

Makaibari and Me

It was nineteen years ago when I first met Rajah in a cafe in Hampstead. A lady at the Tea Board of India who described Makaibari as the 'best champagne you will ever taste' had introduced us, but I was a consultant and knew nothing about teas. Motivated by parenthood and the desire to find healthy foods for my children, I had been harking back to familiar foods from my childhood. Tea was one of them, although as children we were never allowed to drink any because of course 'you must drink your milk first'.

In walked Rajah, resplendent in his bandh gala and speaking nineteen to the dozen. I asked him what made Makaibari different and he said, 'I don't know. Come and see for yourself'. But by this point, I was already there within the tea bushes, listening to the birdsong and the chattering tea pickers. All the while I was thinking about how I could get involved and get the Makaibari story out to people who would love to know it. Life changed dramatically for me. Mission and corporate philosophy suddenly held less magic and my impending MBA at the London Business School became a remote thought. I called around shamelessly telling anyone who would listen about this organic tea estate. It was bound to be an uphill struggle. 'What's organic?' 'Oh come on, all tea is organic.' 'I like the tea I have, thank you.' 'First Flush — is that for symptoms of menopause?' These were the responses. But slowly and surely I was learning about important things — how to tell people things they don't necessarily want to hear, and about a deep sense of concern inside me for our children, our planet and

the roots of our civilisation. I learnt about the human side of business, and also about the not so agreeable corporate side.

The most exciting journey however was as Rajah prophesied — actually being at Makaibari. I made the trip by train from Kolkata. As the day dawned, we passed fields and little villages en route to Kurseong, the nearest town to Makaibari, and then bundled into a car for the last part of the trip. As you take the winding Pankhabari Road up to Makaibari, the story unfolds. Militant tea bushes standing in regimented lines give way to lush forests cradling swathes of tea bushes. Cobwebs sparkle as spiders build furiously, secure in the knowledge that no harmful sprays will ever come their way at Makaibari, and of course they never do. Dig a handful of soil and you wrestle with earthworms, layers of mulch and all the sweat and

in harmony with nature

toil of years of biodynamic husbandry. Walk in the woods and you are intoxicated with the scent of cloves and the heady smell of jasmine, rejoicing in the natural harmony at Makaibari.

And the people; this was the central experience for me. India is unforgiving and life is hard for most people. Daily infrastructural problems, scarcities, poor health care, employment insecurity – the list is endless. Knowing this, I begged Rajah to send me one of his ladies to help me in my home in London with a promise of status, lots of holidays, and more money. He asked around and soon Tea Leaf, as she was nicknamed, was on the train to Mumbai, planning to fly out to London to join me. But this was not to be. 'I want my Rajah sahib. I want my Makaibari,' she wailed. And so Tea Leaf was reunited with her tea bushes, the Makaibari family was complete, and her world was right again.

This is what Makaibari is all about. Rajah's deep commitment and love for the people is returned in spades, as is his strong connection with the land and the plant life. This intense spiritual union among the three has given birth to all of the innovations and developments of unique products and methods of farming. Every trial and tribulation and each success is accompanied by a community response, positive or negative and is thrashed out democratically. Each week, there is a 'darbaar' with Rajah and the women 'councillors', where joys and sorrows are shared, extramarital affairs shamed and plucking standards calibrated. The ladies who run the different sections give their report, and air any concerns, 'Durgi didn't turn up for work yesterday, she hasn't taken a day off for years'. 'Young Mohan has had his eyes operated on and is coming home tomorrow'. 'The panther was sighted this morning'. Families are all provided with a cow so that they can be self-sufficient in milk and biogas. This means trees are not used for firewood and the extra manure is sold to the estate to be used for composting. The forests have been protected in sharp contrast to other areas in India where hills are left with scars where forests once were. Makaibari is where ladybirds breed and Artemisia is grown to ward off the pests of the tea plant – camellia sinensis. They are all in it together and how privileged I am to be a part of this happy world.

Kiran Tawadey

Denna Weiss, is a petite bundle of unbounded positivity. She arrived at Makaibari to live with the community and ended up finding her vocational call – teaching.

My time at Makaibari

As graduation from college came closer, I gave consideration to finding a job and entering the work force – a path laid out by university counsellors and the path that many of my friends had already chosen. What would I do? With my degree in English I could go into a professional field, perhaps publishing, or business writing. None of these options seemed particularly stimulating. So, I began perusing the internet for alternatives. Eventually, I decided to spend sometime in Asia doing volunteer work in an effort to better understand the world and my place in it. So, after much planning, I arrived at the Makaibari Tea Estate in early March. The air was cold and damp, and clouds covered most of the views. Despite the grey-coloured atmosphere, there was something that drew me to this place. I felt an instant connection to the land and the people.

After discussing the options available to me for volunteering my services, I decided to teach at a local village school. On my first day, I walked into another volunteer's classroom to observe. The head teacher came up to me and excitedly asked if I was another new volunteer. I answered yes. The next question determined my fate for the upcoming months when she asked me, 'Do you like babies?' Again, I answered yes. The next thing I knew I was standing in front of a classroom of two, three, four, and five-year-olds who were proudly reciting their ABC's, counting aloud, and singing as many nursery rhymes as they could think of. When their repertoire ended, they all stared at me to see what I knew. So, thinking quickly, I too sang songs and danced around the classroom. That night, I started planning lessons and making goals for the classroom. I also coaxed Mike, my boyfriend and travelling companion, into co-teaching with me.

The weeks passed by and I familiarised myself with Makaibari; making new friends, taking walks through the tea gardens, and slowly learning Nepali phrases to show off to the local people. Before I realised it, Makaibari had become a part of me. It was now my home. I wondered to

myself what was missing here. Everything seemed pretty well in place but there was still a part of me that felt that there was something missing. Then it hit me. There was no library for the community to enjoy. Throughout my life, I have placed a great deal of importance on literacy and have enjoyed spreading my passion for reading with others. What better place to start a library than at Makaibari. I approached Rajah with my suggestion and he whole-heartedly agreed to the plan. Then, Mike and I emailed all our friends and family to tell them the good news, and of course, ask for donations. We were amazed at the amount of support we got from people at home and quickly realised that our initial plan of building a library had already been outgrown. After putting our heads together, Rajah, Mike, and I decided that we would need a new building for the library.

Meanwhile, my plans to stay at Makaibari for three months turned into plans to stay for six months. Currently, the library is being built and books are being purchased and catalogued. When I walk through the villages or down the main street that stretches through Makaibari, I am honoured to be able to smile and casually address my friends and neighbours with a cheerful 'Namaste'. Even though I am thousands of miles from home, I have a family in Makaibari and am a part of this thriving and brilliant community.

Also, up to this point I can thank Makaibari for helping me realise what my future goals are. Now, after returning home, I plan to pursue a degree in Education in order to become a primary school teacher. My love of learning and children are perfectly suited for this career and I can not think of a better way to spend my life. Also, I plan to be involved with NGO work for the rest of my life as it has been a completely life-altering and eye-opening endeavour for me. I thank everyone at Makaibari for taking me in as a sister and opening my eyes to the world around me.

Denna Weiss

Michael Matergia is a serious young man; assured and confident. He had decided to become a doctor. While he was living as a volunteer with the Makaibari community, his daily interactions inspired a clutch of innovative solutions for uplifting the community's health and hygiene.

My time at Makaibari

To self-reflect on Makaibari is nearly an impossible task. Swirling through my mind are the sounds of Halderkothi waking up to a new day, children climbing the hill on their way to school, butterflies floating through the air, the smells of a cardamom rich curry, and hot, steaming tea all ready to be enjoyed, or perhaps of the flavor trapped in the nascent shoots of the bushes which cling to hills, up and down the valley.

Denna with school children

More and more memories rush through my head and are hard to grab hold of as another thought takes hold before I have a chance to jot down the first. Yangi Lama sounding out the words to Dr Seuss's 'Hop on Pop', Mohin in contemplation, Aitu simmering a dish, Rajah riding through the field, Shila curing the community, and the nursery class enjoying itself. I could go on for hours, days, and many pages.

I have been told that Makaibari translates as 'corn-field' but for me it means something different. When Makaibari passes from the lips of my friends, I hear 'welcome home'. For despite being a thousand miles away from my life in America, I feel and have felt since the day I arrived at Makaibari, completely welcomed into the community. It is a feeling that increases as the days go by. The land, the people, and the life that consummates this special place seeps into my being.

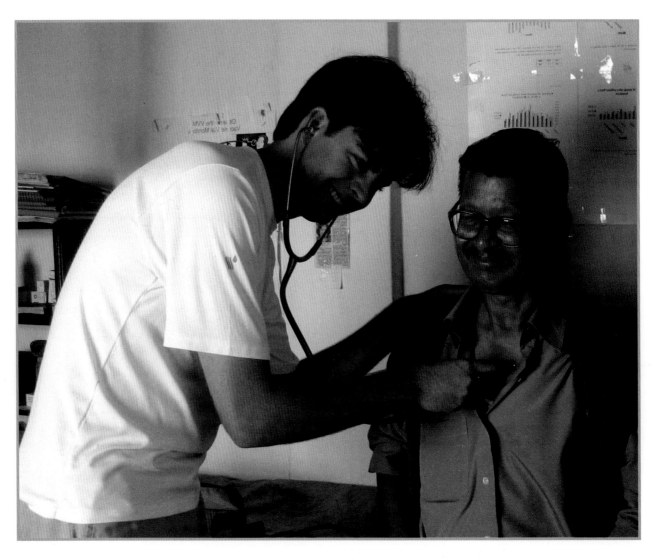

Michael working at the dispensary

The six months at Makaibari have been among the most fulfilling of my young life. Whether I am standing in front of the twenty-one kids in our class at Upper Makaibari or sifting through the health data from the dispensary, I constantly feel the urge to work hard to give something back to a community which has given me so much. At the back of my mind, there is the constant reminder that soon I will have to leave. When I depart though, I will take with me all the beautiful memories of the life long friends that I have made and the children that I have taught. It is my sincere wish that one day I will be able to return; to hear once again the word 'Makaibari' and know that I am coming home.

Michael Matergia

Makaibari villagers selling their wares at Kurseong

Epilogue

the Rashtriya Rattan Award

On 21 November, 2002, the Global Economic Council bestowed upon me the Rashtriya Rattan Award for Makaibari. 'Rashtriya Rattan' means jewel of the nation, and the award is given for outstanding individual achievement and distinguished services to the nation. It symbolises unity (ekta) achieved by creative (nirman) private enterprise (udyog) that glorifies (gaurav).

In July 2003, Makaibari's Silver Tips Imperial fetched a world record price at the Kolkata auctions conducted by J. Thomas and company at Rs. 18,000/- per kg.

Xavier de Lauzanne, a renowned documentary filmmaker from France, has made an award winning documentary *The Lord of Darjeeling*. This hour-long documentary (included with this book) captures the true spirit of Makaibari and is moving audiences at various film festivals globally.

Discovery channel and Travel and Living, featured Makaibari as the world's best tea in May/June /July 2007 in 'A Matter of Taste – Two Leaves and a Bud'.

Since early 2006, the BBC had been interviewing sixty Indians selectively. The interviews resulted in a four part serial, aired around 15 August, 2007 when India celebrated her sixtieth year of independence. The purpose of the interviews was to showcase the sixty Indians who are likely to impact India's next sixty years dynamically. I was one of the sixty selected in this honour roll.

looking towards a new dawn

Sanjeev Bhaskar, the celebrated star of British TV was the anchor who interviewed me at work and play at Makaibari. He was positively moved by his time spent at Makaibari and was highly appreciative of the sustainable impulses at play here. The vision of Makaibari has been extended beyond its boundaries.

The movement of 'Organic Ekta' is a Makaibari joint venture with CHAI and Darjeeling Earth Group. The two NGOs work with a continually growing list of small farmers of the Darjeeling region and monitor the growing of herbs, fruits, vegetables, cereals and tea organically. The use of biogas as a fuel alternative has captured the imagination of these growers – which portends considerable hope for greening the Darjeeling region to its pristine splendour of yore. Coupled with numerous innovative marketing impulses, a new era of sustainable entrepreneurship is witnessing its dawn – for a radiant future.

Index

A
Afforestation, 16, 64
    Masanobu Fukuoka's technique, 53–55
    schemes by female workers, 16–17
agriculture, in context of human evolution, 24–25
Assamese Macaque, 113
Awards, 164
Aziz, Tariq, 37–39

B
Bai Mu Dan, 75
bamboo, uses of, 17, 31
Banerjee, Girish Chandra (G.C.), 2–3
banji period, 71–73
Bawanipokri Sanctuary, 115
Bhattacharya, A.K., 76
biodynamics, 25–26, 28, 31, 35–36
BioFach, 97
biogas energy, 16, 21, 22–23, 28, 101–102, 165

C
Campbell, Dr, 2
CHAI, 165
chamomile flowers, 30
Chatt, Rachida, 105
chemical poisoning and impacts, at tea plantations, 14
clay pellets, 54–55
cobra, 21, 133
Coffee Khety, 111
cow pat manure, 46
Crees, 5–6

D
Dandelion flowers, 30
Darjeeling, 2–3, 51–57, 72, 75
Darjeeling Earth Group, 165
Darjeeling vintages, 45, 71
de Lauzanne, Xavier, 164
Demeter certificate, 31
Demeter-International, 97
Dia Tea, 76
Dussehra, 120–121

E
Earthworms, 49
Eco agriculture, 98
elephants, 114–115

F
Fairtrade Labelling Organisation (FLO), 65, 95, 98, 106
Fairtrade sales, 103–105
Faltier, Gunther, 151–154
female workers, 15
    and innovative afforestation scheme, 16–17
Fermentation process, 71
First Flush tea shoots, 17, 21, 45–46, 71, 77
FTGOFPIS CHINA tea, 46
Fukuoka, Masanobu, 53–54, 83

G
Gandhi, 15–16
Golden Langur, 112
Green tea, 74–77
Gyokuro Green Tea, 99–100

H
1978, 2002 hailstorm, impacts, 17–18, 46
the Hampstead Tea and Coffee Company, 23–24, 97
herb preparations, 29–30
holi festival, 45–46
homeopathic preparation
    BD 500, 29
    BD 501, 29
    BD 502, 30
    BD 508, 29–30
Honma, Yuko, 54–55

I
Indian Pied Hornbills, 111–112
Indian Plantations Labour Act, 106
Indian undergraduate's existence, at the University of London, 6–7
Ishii family, 47
Ishii, Yoshihiri, *see* Toyota Tiger

J
Japanese Agricultural Systems Society (JASS), 47–48

K
Kapukapu restaurant, 100
khada, 54, 89, 140
Kikuchi, K., 82
Kodobari village, 16, 121
Kurseong, 2–3, 50
Kurseong Hospital, 56